Hydraulics of Levee Overtopping

IAHR Monograph

Series Editor: Robert Ettema
Department of Civil and Environmental Engineering,
Colorado State,
Fort Collins,
USA

The International Association for Hydro-Environment Engineering and Research (IAHR), founded in 1935, is a worldwide independent organisation of engineers and water specialists working in fields related to hydraulics and its practical application. Activities range from river and maritime hydraulics to water resources development and eco-hydraulics, through to ice engineering, hydroinformatics and continuing education and training. IAHR stimulates and promotes both research and its application, and, by doing so, strives to contribute to sustainable development, the optimisation of world water resources management and industrial flow processes. IAHR accomplishes its goals by a wide variety of member activities including: the establishment of working groups, congresses, specialty conferences, workshops, short courses; the commissioning and publication of journals, monographs and edited conference proceedings; involvement in international programmes such as UNESCO, WMO, IDNDR, GWP, ICSU, The World Water Forum; and by co-operation with other water-related (inter) national organisations. www.iahr.org

Supported by
**Spain Water
and IWHR, China**

Hydraulics of Levee Overtopping

Lin Li

Department of Civil and Architectural Engineering, Tennessee State University, TN, USA

Farshad Amini

Department of Civil and Environmental Engineering, Jackson State University, MS, USA

Yi Pan

College of Harbor, Coastal and Offshore Engineering, Hohai University, Nanjing, China

Saiyu Yuan

State Key Laboratory of Hydrology – Water Resources and Hydraulic Engineering, Hohai University, Nanjing, China

Bora Cetin

Department of Civil, Construction and Environmental Engineering, Michigan State University, MI, USA

CRC Press
Taylor & Francis Group
Boca Raton London New York

CRC Press is an imprint of the
Taylor & Francis Group, an **informa** business

CRC Press/Balkema is an imprint of the Taylor & Francis Group,
an informa business

© 2021 Taylor & Francis Group, London, UK

Typeset in Times New Roman
by codeMantra

Library of Congress Cataloging-in-Publication Data

Applied for

Published by: CRC Press/Balkema
                  Schipholweg 107C, 2316 XC Leiden, The Netherlands
                  e-mail: Pub.NL@taylorandfrancis.com
                  www.routledge.com – www.taylorandfrancis.com

ISBN: 978-0-367-27727-7 (Hbk)
ISBN: 978-0-367-53507-0 (pbk)
ISBN: 978-0-429-29755-7 (eBook)
DOI: 10.1201/9780429297557
DOI: https://doi.org/10.1201/9780429297557

About the IAHR Book Series

An important function of any large international organisation representing the research, educational and practical components of its wide and varied membership is to disseminate the best elements of its discipline through learned works, specialized research publications and timely reviews. IAHR is particularly well-served in this regard by its flagship journals and by the extensive and wide body of substantive historical and reflective books that have been published through its auspices over the years. The IAHR Book Series is an initiative of IAHR, in partnership with CRC Press/Balkema – Taylor & Francis Group, aimed at presenting the state-of-the-art in themes relating to all areas of hydro-environment engineering and research.

The Book Series will assist researchers and professionals by advancing and transferring contemporary knowledge needed for research, education and engineering practice. The series includes Design Manuals and Monographs. The Design Manuals, usually prepared by multiple authors, guide the application of theory and research findings to engineering practice; and, the Monographs give state-of-the-art coverage of various significant topics in water engineering.

The first and highly successful IAHR book was "Turbulence Models and their Application in Hydraulics" by W. Rodi, published in 1984 by Balkema. "Turbulence in Open Channel Flows" by I. Nezu and H. Nakagawa, also published by Balkema (in 1993), had an important impact on the field and, during the period 2000–2010, further authoritative texts (published directly by IAHR) included Fluvial Hydraulics by S. Yalin and A. Da Silva and Hydraulicians in Europe by W. Hager. All of these publications continue to strengthen the reach of IAHR and to serve as important intellectual reference points for the Association.

Since 2011, the Book Series is once again a partnership between CRC Press/Balkema – Taylor & Francis Group and the Technical Committees of IAHR. The present book is an exciting further contribution to IAHR's Book Series, substantially aiding water engineering research, education and practice, and showcasing the expertise IAHR fosters.

Series editors
Robert Ettema

Contents

Preface

The critical role of effective earthen levees to protect populations and infrastructure from storms and hurricanes is widely acknowledged by all involved in the design, construction, operation, and maintenance of levees. The strong weather systems, as a result of climate change, have resulted in sea level rising at an increased rate, as well as storms of increased intensity and duration. Overtopping may occur during the periods of flood due to insufficient freeboard. Overtopping of levees produces fast-flowing, turbulent water velocities on the landward-side slope that can damage the levee. It includes surge-only overflow, wave-only overtopping, and combined wave and surge overtopping. If overtopping continues long enough, the erosion may eventually result in loss of levee crest elevation and perhaps breaching of the protective structure. The catastrophic consequences of levee overtopping have been demonstrated during many hurricanes, such as Hurricane Katrina and Sandy

This book focuses on the effectiveness of three innovative levee-strengthening systems during true full-scale overtopping conditions simulating surge, waves, or combined wave and storm surge: high-performance turf reinforcement mats (HPTRM), articulated concrete block (ACB) (hard-armor products), and roller-compacted concrete (RCC). Overtopping volumes of water are highly turbulent and have substantial amount of air entrainment. Little is known about the more problematic and complex case of nonstationary high turbulent flow conditions. These turbulent flows significantly impact the design parameters for the various strengthening methods. This book also presents the impact of the turbulent flow on the performance of HPTRM-strengthened levee during true full-scale overtopping conditions. Extensive numerical modeling as well as the development of design and construction guidelines for an innovative and cost-effective protection system are also presented. This book only covers the hydraulic performance of levee systems, and thus, the geotechnical slope stability of various strengthened levee systems under combined wave and surge turbulent overtopping conditions is not discussed in this book. The geotechnical slope stability of these systems under various conditions have been reported by the authors in many publications.

This book originates from two major research projects conducted by the authors (Farshad Amini and Lin Li) for the Department of Homeland Security-sponsored Southeast Region Research Initiative (SERRI) at the Department of Energy's Oak Ridge National Laboratory, beginning in 2009. Obviously, the views and conclusions contained in this book are those of the authors and should not be interpreted as necessarily representing the official policies, either expressed or implied, of the U.S.

Department of Homeland Security. Mention of specific products or trade names is for informational purposes only, and does not constitute an endorsement or recommendation for their use. The full-scale hydraulics laboratory experiments were conducted by O.H. Hinsdale Wave Research Laboratory at Oregon State University. The soil erodibility of strengthening materials was tested by Texas Transportation Institute at the Texas A & M University. The mechanical properties of three levee-strengthening materials were tested by Mechanical Characterization and Analysis User Center at Oak Ridge National Laboratory. Thanks are also extended to our students, including Clifton Hulitt, Justin Roberts, William McCleave, Fabio Santos, Xin Rao, Jianhua Wu, Yi Pan, Saiyu Yuan, and Wei Shao. We also thank Richard Goodrum of Colbond, Inc., Michael Robenson, Kevin Spittle, Jessie Clark of Profile Products, Inc., Dr. Tim Maddux, Dr. Daniel Cox, and Jason Killian of the Oregon State University, and Dr. Amit Shyam and Dr. Edgar Lara-Curzio of the Oak Ridge National Laboratory for their support during the laboratory experiments. We also thank Dr. Norberto Nadal of the U.S. Army Corps of Engineers, Engineer Research and Development Center (ERDC) for peer-reviewing the final report.

Our sincere appreciation is extended to all involved in helping us develop this book.

Lin Li, PhD, PE, Fellow of ASCE **Farshad Amini**, PhD, PE, Fellow of ASCE
Tennessee State University; Jackson State University
Nashville, TN, USA; Jackson, MS, USA

Variables

a	Scale factor of Weibull distribution
b	Shape factor of Weibull distribution
C_f	Empirical friction of surge-only overflow discharge
d_m	Average flow thickness on landside slope
d_s	Steady flow thickness on landside slope
Emax	Maximum soil loss depth
f_F	Fanning friction factor
f_{F*}	Equivalent Fanning friction factor
g	Gravity
$h$rms	Root-mean-square flow thickness
h_1	Upstream head (difference between surge elevation and levee crest elevation, $h_1 = -Rc$)
$h_1/3$	Average flow thickness of the highest $1/3^{rd}$ of the peak flow thicknesses
$h_1/10$	Average flow thickness of the highest $1/10^{th}$ of the peak flow thicknesses
$h_1/100$	Average flow thickness of the highest $1/100^{th}$ of the peak flow thicknesses
$Hm0$	Energy-based significant wave height
Hrms	Root-mean-square wave height
H_s	Significant wave height
$H_1/3$	Average height of the highest $1/3^{th}$ of the individual waves
$H_1/10$	Average height of the highest $1/10^{th}$ of the individual waves
$H_1/100$	Average height of the highest $1/100^{th}$ of the individual waves
k_d	Empirical parameter related to flow thickness of landside slope
k_{dm}	Empirical parameter related to average flow thickness of landside slope
k_v	Empirical coefficient of steady flow velocity on landside slope
$L0$	Deepwater wavelength
L_p	Deepwater wavelength based on peak period
n	Manning coefficients
P_{ow}	Probability of overtopping
q_s	Steady overflow discharge
q_w	Average overtopping discharge
q_{ws}	Average combined wave and surge overtopping discharge
R_c	Freeboard
R_e	Reynolds number
T_m	Mean period
$T_m-1,0$	Mean (energy) wave period

T_p	Peak period
$u*$	Friction velocity
v_c	Critical overflow velocity
v_{s*}	Steady flow velocity on landside slope
v_m	Mean flow velocity on landside slope
v_w	Wave front velocity
Vmax	Maximum individual wave volume
Vmean	Average wave volume
α	Slope gradient
β	Landside slope angle
$\xi_m{-}1{,}0$	Breaking parameter based on deepwater wavelength and mean energy period
ξ_p	Breaking parameter based on deepwater wavelength and peak period
ρ	Water density
τ	Shear stress
τ_m	Average shear stress on landside slope
τ_s	Shear stress on landside slope

Chapter 1

Introduction

1.1 Background

Storm surge is a disastrous natural phenomenon referring to the abnormal rise in sea level caused by strong weather systems such as typhoons, extratropical cyclones, strong wind action of cold fronts, and sudden changes in barometric pressure. Storm surge causes the tide level in the affected sea area to significantly exceed the normal tide level. If the storm surge occurs simultaneously with the spring tide or upstream floods, it can lead to catastrophic disasters.

Over the past decades, the number and cost of disaster events in the United States have risen significantly (NOAA 2019), in particular, for hurricanes. Hurricanes Katrina in2005, Sandy in2012, Harvey in 2017, Maria in2017, Irma in 2017, and Michael 2018 caused the most damage in the United States. Hurricane Katrina in August 2005 was one of the most devastating and deadliest disasters in U.S. natural disaster history with an estimated loss of 1,200 human lives, with 500,000 people having left the area and 125 billion dollars of economic loss (Scawthorn and Porter 2019). Hurricane Sandy (2012) affected 24 states, with particularly severe damage in New York City due to storm surge, which flooded streets, tunnels, and subway lines and cut power in and around the city. Hurricane Harvey in 2017 brought unprecedented rainfall, basin-wide flooding, and windstorms that devastated infrastructure and flooded more than 150,000 homes. It approximately matched the damage of Katrina as the costliest hurricane on record, with $125 billion in damages (NHC 2018), primarily due to flooding in Houston and Southeast Texas. Hurricane Maria (2017) caused over 3,000 deaths and extensive damage to Puerto Rico. Hurricane Michael, which devastated northwest Florida in 2018, was the strongest storm on record for southeast U.S. and led to an estimated loss $25 billion (Baecher et al. 2019).

Cities and areas in these regions were designed with earthen levees, which are one of the most common structures built to protect against flooding and extreme event-related disasters. Earthen levees are used extensively in the United States to protect populations and infrastructure from periodic floods and high water levels caused by storm surges. There are approximately 13,679 km of levees in the United States. The causes of failure for levees include overtopping, surface erosion, internal erosion (piping), and slope instability (USACE 2000; Perry 1998; TAW 2002; ASCE 2011). Climate change is well-known to result in sea level rising at an increased rate and causing storms of increasing intensity and duration (IPCC 2012). Overtopping can occur during flooding caused by insufficient freeboard. Overtopping of levees produces fast-flowing, turbulent water velocities on the landward-side slope that can damage the protective grass covering and expose the underlying soil to erosion. If overtopping continues long enough, the erosion may eventually result in loss of levee crest elevation and ultimately breaching of the protective structure. The catastrophic consequences of levee overtopping were seen during Hurricane Katrina in the United States in August 2005. The levees were subjected to a catastrophic 8.23 m estimated storm surge during Hurricane Katrina (Grenzeback and Lukmann 2007). The surge severely strained the levee system in the New Orleans area, which began to fail as early as the morning of August 29, 2005. The levees in the New Orleans area breached at about 50 distinct locations. At least seven of the major failures were related to breaching of levees containing I-walls. The rest of the levees breached when they were overtopped by floodwater, which eroded the levee material away. Figure 1.1 shows the flow over the crest and the erosion of the landward-side slope of levees during Hurricane Katrina.

Overtopping associated with earthen levees includes surge-only overflow, wave-only overtopping, and combined wave and surge overtopping. Surge-only overflow occurs when the surge level exceeds the levee crest elevation without accompanying wave action. Wave-only overtopping occurs when the surge level is below or equal to the levee crest elevation. The wave-only overtopping is unsteady compared to the steady surge-only overflow. After passing over the levee crest, each overtopping wave has a triangular discharge distribution with a maximum discharge at the leading edge that is several times greater than the time-averaged discharge. The most severe overtopping condition is when the levee is being overtopped by combined waves and surge (Hughes and Nadal 2009).

Increased rates of sea level rising and storms of increased intensity and duration (IPCC 2012; Baecher et al. 2019) increase the risks of wave overtopping under negative freeboard. Since Hurricane Katrina, wave overtopping under negative freeboard has been one of the main focus of studies. Post-Katrina investigations revealed that most earthen levee damage occurred on the levee crest and landward-side slope as a result of either wave overtopping, storm surge overflow, or a combination of both (ASCE Hurricane Katrina External Review Panel 2007). According to Hughes and Nadal (2009), during the combined wave and surge overtopping, the overtopping waves can create critical flow conditions near the leeward edge of the levee crest, resulting in

Figure 1.1 Flow over the crest (a) and the erosion of the landward-side slope of levees (b) during Hurricane Katrina. (source: http://www.recmod.com/.)

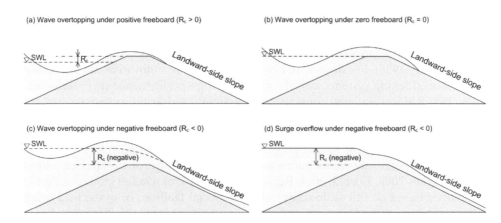

Figure 1.2 Wave overtopping under (a) positive, (b) zero, (c) negative freeboard, and (d) surge overflow. (Adopted from Pan et al. (2015a). Reproduced with permission from Elsevier.)

supercritical wave overtopping flow on the landward-side slope. Figure 1.2 illustrates the wave overtopping under positive, zero, and negative freeboard and surge overflow.

Wave-only overtopping and surge-only overflow are also classical problems in the coastal engineering field. Several studies have been conducted on these topics, including physical experiments and numerical modeling for the design of levees.

Levees in soft soil or loose sediment areas often subside continuously, which results in the reduction of flood control standards. These two factors increase the risk of combined wave and surge overtopping. In recent years, several serious erosion and damage events have been observed on the landward-side slope of levees caused by combined wave and surge overtopping, which has made combined wave and surge overtopping an important factor in the design of levees. Therefore, predicting the effects of combined wave and surge overtopping on levees under extreme conditions has become one of the urgent problems to be solved in the field of coastal disaster prevention.

According to previous analyses, during overtopping, the landward-side slope of levees is subjected to significantly higher velocities and much greater erosive forces than the flood-side slope (Hughes and Nadal 2009; Li et al. 2012; Pan et al. 2013a, 2013b; Yuan et al. 2015a, 2015b; Pan et al. 2016). Field tests also indicate that erosion failure occurs first on the landward-side slope of the overtopped levee and progressively regresses (Hanson et al. 2003, 2005; van der Meer et al. 2009). Hence, protecting levees from erosion by surge overflow and wave overtopping is necessary to assure a viable and safe levee system (van der Meer et al. 2002; Sills et al. 2008; Pan et al. 2015a, 2015b; Baecher et al. 2019).

An evaluation of the various innovative overtopping protection methods indicates that a cellular confinement system using concrete-filled cells would be applicable for use when rapid, rather expensive, construction is justified and the subsequent rising of the levee is anticipated. Flow velocities in the range of 1.8–3.0 m/s can be handled using this approach. Reinforced grass would have limited application because it would take several months (or years) for the root system to develop and it would be limited (unless more flood-resistant grass and/or additional anchorage is provided) to flow durations less than 2 days. In addition, reinforced grass can handle flow velocities up to 3.8 m/s. Although reinforced grass is generally more economical than conventional engineering materials in capital cost, it can be more expensive to maintain. Soil cement is, in general, cost-competitive for nonsediment-laden flows with velocities up to 7.6 m/s (Perry 1998). For flows with high velocities, three innovative and cost-effective levee-strengthening systems, namely, anchored high-performance turf reinforcement mat (HPTRM), articulated concrete block system (hard-armor products), and roller-compacted concrete (RCC) system may be considered.

The erosion process of levees is difficult to model numerically or in small-scale physical models, and the roughness of strengthening systems must be determined by full-scale testing in laboratory flumes using defined testing protocols (Akkerman et al. 2007; Hughes 2008; Li et al. 2014; Pan et al. 2015a, 2015b). Design guidance should be developed based on full-scale study either in special facilities or in the field using an apparatus such as the Overtopping Simulator (van der Meer et al. 2009; Baecher et al. 2019). However, the present design criteria only include guidance for both grass slopes and a variety of slope protection products for steady overflow. This book is directed toward developing an innovative and cost-effective levee protection system during overtopping.

1.2 Contents of this book

The land sides of levees are subjected to significantly higher velocities and much greater erosive forces than the flood side of the levees during wave overtopping, surge overflow, or combined wave and surge overtopping in a flood. Robust levee reinforcement is critically needed to be installed on the land sides of levees to resist erosion damage. Previous research and tests have focused almost exclusively on the conditions of steady-state overflow. Therefore, little is known regarding the more problematic cases of unsteady overtopping caused by waves only or combined wave and storm surge.

The combined wave and surge overtopping has attracted more interest in the field of coastal engineering since 2005. This book describes a physical model study of a full-scale (1:1) model of levee overtopping by combined wave and surge overtopping conducted at Oregon State University (OSU). The purpose of this book is to study the hydrodynamic characteristics of the combined wave and surge overtopping process and to study the erosion characteristics of the landward-side slope of levees with different strengthening systems under combined wave and surge overtopping. The three strengthening methods include the use of anchored HPTRM, articulating concrete block system (hard-armor products), and RCC system. The tested levee cross-section was built in the 4.572 m deep wave flume. The levee model was built with a sand core and a concrete cap with a 0.76 m deep and 2.34 m wide test section, which was used to install different protection layers. Eight surge-only overflow tests under different upstream water levels and 24 combined wave and surge overtopping tests under different combinations of incident waves and average sea level were conducted. Hydrodynamic parameters and erosion characteristics in the test section of the landward-side slope of the levee model were recorded. This study is the first in the world to use a scale of 1:1 to investigate the combined wave and surge overtopping, representing true full-scale conditions.

This book presents the experimental settings, data analyses, and research conclusions in detail. Chapter 1 is the introduction; Chapter 2 introduces the existing studies of surge-only overflow and overtopping as the background of combined wave and surge overtopping research; Chapter 3 discusses the three levee-strengthening systems and their engineering properties; Chapter 4 introduces test sections constructions, instrumentation, testing procedures, and initial data analysis of large-scale wave flume test. The scale effects and the model and measurement effects are briefly discussed in Chapter 4; Chapter 5 introduces the sampling and erosion function apparatus tests of HPTRM- strengthened levee after large-scale wave flume test; Chapter 6 presents the combined wave and surge overtopping process and the calculation method of relevant hydraulic parameters through the analysis of large-scale flume test data. Chapter 6 analyzes the equivalence between surge-only overflow and combined wave and surge overtopping. A general method for estimating the hydraulic parameters of combined wave and surge overtopping is proposed. Chapter 7 presents the turbulent flow analysis for the overtopping hydraulic tests. Overtopping measurements of near-bottom velocity were provided to produce direct covariance estimates of turbulent shear stress. In Chapter 8, the erosion characteristics of three kinds of levee protection layers are analyzed. On this basis, the conceptual model of landward-side slope erosion of levees and the testing method of HPTRM-strengthened levee were proposed. Limitations of physical scale or experimental equipment made it difficult to observe the supercritical overtopping flow at the toe of the landward-side slope. Numerical models were often employed to fill the gaps that physical models have due to limitations of instruments in the wave overtopping studies. Chapter 9 discusses two numerical methods to study the overtopping hydraulics at the toe of the landward-side slope of HPTRM- and RCC-strengthened levee under combined wave overtopping and storm surge overflow. Chapter 10 presents a three-dimensional hydrodynamic and sediment transport model, ECOMSED, to simulate overtopping hydrodynamic flow, turbulent shear stress, turbulent kinetic energy, and erosion rate at the toe of landward-side slope under combined overtopping condition.

Chapter 2

Surge overflow, wave overtopping, and combination

The research results of surge-only overflow and overtopping on levees are the foundation of combined wave and surge overtopping research, as well as the main issues in the coastal engineering field. The key parameters of surge-only overflow include discharge, overflow velocity, and stress. On the other hand, the key parameters of overtopping include average overtopping discharge, overtopping distribution, and overtopping probability. To understand the characteristics of combined wave and surge overtopping, this chapter provides a review of the existing research on surge-only overflow and overtopping.

Because of the catastrophic consequences of Hurricane Katrina, the research on combined wave and surge overtopping has attracted the attention of scholars and researchers. To estimate hydraulic parameter as soon as possible, some emergency researches have been conducted by scholars from all over the world. This chapter introduces the existing research results on combined wave and surge overtopping.

2.1 Surge overflow

For a steady water overflow of a levee caused by a storm surge higher than the levee crest (i.e., negative freeboard with $R_c < 0$), subcritical flow exists on the high-water side of the levee. If the horizontal levee crest is sufficiently long to maintain a hydrostatic pressure distribution, critical flow (transition between subcritical and supercritical flow) occurs somewhere along the levee crest; moreover, the flow down the landward-side slope is supercritical unless the slope is very mild.

2.1.1 Surge overflow discharge

When the steady water level remains unchanged, the overflow also maintains a steady flow. Assuming minimal frictional energy losses along the crest, the discharge per unit length of the levee is computed by the generally accepted equation (2.1) for flow over a broad-crested weir given by open channel flow texts as:

$$q_s = \left(\frac{2}{3}\right)^{3/2} \sqrt{g} h_1^{3/2} \tag{2.1}$$

where q_s is the steady overflow discharge per unit length, g is gravity acceleration, and h_1 is the upstream head (the difference between surge elevation and levee crest elevation, $h_1 = -R_c$).

According to Kindsvater (1964), when the steady water level remains unchanged, the overflow will also maintain a steady flow. Assuming minimal frictional energy losses along the crest, the discharge per unit length of the levee is computed by the generally accepted equation (2.2) for flow over a broad-crested weir given by open channel flow texts as:

$$q_s = C_f \sqrt{g} h_1^{3/2} \tag{2.2}$$

where q_s is the steady overflow discharge per unit length, g is gravity acceleration, and h_1 is the upstream head (the difference between surge elevation and levee crest elevation, $h_1 = -R_c$).

2.1.2 Critical water depth and velocity

When the upstream water level remains unchanged, the overflow also maintains a steady flow. Assuming minimal frictional energy losses along the crest, the discharge per unit length of levee is computed by the generally accepted equation (2.3) for flow over a broad-crested weir given by open channel flow texts as:

$$q = \sqrt{gh^3} \, F_R \tag{2.3}$$

where q is the steady overflow discharge, h is the flow thickness perpendicular to the slope, and F_R is the Froude number.

For critical flow, $F_R = 1$, Equation (2.3) can be expressed as:

$$q_c = \sqrt{gh_c^3} \tag{2.4}$$

where q_c is the critical overflow discharge and h_c is the critical flow thickness perpendicular to the slope.

If the q_s in Equation (2.1) and the q_c in Equation (2.4) are equal, the following equation can be obtained:

$$h_c = \frac{2}{3} h_1 \tag{2.5}$$

Its velocity can be expressed as:

$$v_c = \sqrt{gh_c} \tag{2.6}$$

where v_c is the critical overflow velocity.

2.1.3 Shear stress of surge overflow

In this book, the time series of shear stress at measuring points along the landward-side slope were calculated using the momentum equation of Saint–Venant equations (2.7) given by:

$$\frac{\partial v}{\partial t} + v \frac{\partial v}{\partial s} + g \frac{\partial h}{\partial s} + gS_f - g\sin\beta = 0 \tag{2.7}$$

where v is the flow velocity parallel to the slope, t is the time, s is the down-slope coordinate, g is gravity acceleration, h is the flow thickness perpendicular to the slope, S_f is the friction slope, and β is the angle of levee slope to horizontal. Equation (2.7) can be rewritten as:

$$S_f = \sin\beta - \frac{\partial h}{\partial s} - \frac{\partial}{\partial s}\left(\frac{v^2}{2g}\right) - \frac{\partial v}{\partial t}\frac{1}{g} \qquad (2.8)$$

The friction slope at any measuring point can be calculated based on Equation (2.8), and then the overflow shear stress can be calculated according to Equation (2.9).

$$\tau_s = \gamma_w h S_f \qquad (2.9)$$

where τ_s is the shear stress and γ_w is the specific weight of water.

2.2 Wave overtopping

Wave overtopping is a common phenomenon during storm surge. Overtopping occurs when waves run up over crests of levees. Levees often fail due to wave overtopping and a failure of the landward-side slope.

2.2.1 Processes of overtopping

To study the interaction between waves and levees, Schüttrumpf and Oumeraci (2005) divided overtopping into five processes, as shown in Figure 2.1. These processes include: (1) wave parameters at the toe of the levee; (2) wave transformation on the levee slope up to the breaking point; (3) wave run-up and wave run-down on the levee slope; (4) wave overtopping on the levee crest; and (5) wave overtopping on the landward-side slope. There are different ways of action between waves and levees during these five processes. A thorough understanding of the hydrodynamic characteristics of each process can help to further understand the erosion and destruction of levees by waves.

Through theoretical analyses, Schüttrumpf and Oumeraci (2005) described the change in the overtopping flow velocity along the crest and the landward-side slope of levees according to Naiver–Stokes equations. As this book focuses on the change in the

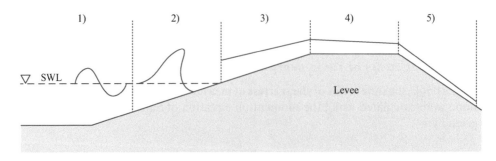

Figure 2.1 Processes of overtopping. (Modified from Schüttrumpf and Oumeraci (2005).)

overtopping flow velocity along the landward-side slope, only the expressions of the velocity and flow thickness along the landward-side slope are provided. The flow velocity on the landward-side slope is given by:

$$v = \frac{v_0 + \dfrac{k_1 h}{f} \tan h\left(\dfrac{k_1 t}{2}\right)}{1 + \dfrac{f v_0}{h k_1} \tan h\left(\dfrac{k_1 t}{2}\right)} \tag{2.10}$$

with

$$h = \frac{v_0 h_0}{v} \tag{2.11}$$

where v_0 is the overtopping velocity at the beginning of the landward-side slope, h_0 is the flow thickness at the beginning of the landward-side slope, f is the bottom friction coefficient, and K_1 and t can be described as follows:

$$k_1 = \sqrt{\frac{2fg \sin\beta}{h}} \tag{2.12}$$

$$t = -\frac{v_0}{g \sin\beta} + \sqrt{\frac{v^2}{g^2 \sin^2\beta} + \frac{2s}{g \sin\beta}} \tag{2.13}$$

Simultaneous Equations (2.10–2.13) and definite boundary conditions (flow parameters at the junction of the crest and landward-side slope) can be used to calculate the velocity and flow thickness at any time and any point on the landward-side slope.

2.2.2 Average overtopping discharge

The wave overtopping parameters are the key parameters in the design of levees and the management of coastal protection. The most representative overtopping parameter is the average overtopping discharge q_w, which is used in the design of the crest level of levees.

Owen (1980) reported an examination of the overtopping discharge, and provided the relationship between dimensionless average overtopping (Q_o) and dimensionless freeboard (R_o) of trapezoidal impervious levees as follows:

$$Q_O = \frac{q_w}{g H_s T_{m0}} = a_O \exp(-b_O R_O) \tag{2.14}$$

with

$$R_O = \frac{R_c}{T_{m0}\sqrt{gH_s}} \tag{2.15}$$

where a_O and b_O are empirical parameters related to the shape of levees profile, q_w is the average overtopping discharge, H_s is the significant wave height, and T_{m0} is the mean (energy) period. The application range of Equations (2.14) and (2.15) is $0.05 < R_O < 0.3$.

Based on the experimental data from Hedges et al. (1998) established a regression model to estimate the average overtopping. The model considers two constraints: (1) when the elevation of the levee crest is very high, the overtopping is zero; and (2) when the freeboard of the levee is zero, the overtopping is a large finite value. The empirical formula obtained can be expressed as:

$$Q_H = \frac{q_w}{\sqrt{g(CH_s)^3}} = A\left(1 - \frac{R_c}{CH_s}\right)^B \tag{2.16}$$

where Q_H is the dimensionless overtopping discharge, A and B are regression coefficients, C is the ratio of maximum wave run-up to incident significant wave height.

Ward and Ahrens (1992) researched overtopping based on a large number of available experimental data. The overtopping test data were divided into seven groups according to the incident wave conditions and the shape of the levee profile. The regression analysis was conducted, and the formula for calculating the average overtopping of the levee with a trapezoidal cross-section was given as follows:

$$\frac{q_w}{\sqrt{gH_{m0}^3}} = C_0 \exp\left[\frac{C_1 R_c}{\left(H_{m0}^3 L_0\right)^{1/3}}\right] \exp\left(C_2 m\right) \tag{2.17}$$

where H_{m0} is the energy-based significant wave height, L_0 is deepwater wavelength, m is the reciprocal of the slope gradient, and C_0, C_1, C_2 are empirical parameters, which depend on the incident wave conditions and the shape of the levee profile.

The formulas of Van der Meer and Janssen (1994) are widely used in engineering design. Van der Meer and Janssen used the breaking parameter to derive the following overtopping formulas for breaking and nonbreaking waves, which described the average overtopping discharge:

For breaking waves ($\xi_p < 2$):

$$Q_v = \frac{q_w}{\sqrt{gH_s^3}}\frac{\sqrt{\tan\alpha}}{\xi_p} = 0.06\exp\left(-5.2\frac{R_c}{H_s\xi_p}\frac{1}{\gamma_r\gamma_b\gamma_h\gamma_\beta}\right) \tag{2.18}$$

For nonbreaking waves ($\xi_p > 2$):

$$Q_v = \frac{q_w}{\sqrt{gH_s^3}} = 0.2\exp\left(-2.6\frac{R_c}{H_s}\frac{1}{\gamma_r\gamma_b\gamma_h\gamma_\beta}\right) \tag{2.19}$$

where Q_v is the dimensionless overtopping discharge, α is the slope gradient, γ_r is the reduction factor for friction, γ_b is the reduction factor for berm, γ_h is the reduction factor for water depth, γ_β is the reduction factor for incident wave angle, and ξ_p is described as:

$$\xi_p = \frac{\tan\alpha}{\sqrt{H_s / L_p}} \tag{2.20}$$

where L_p is described as:

$$L_p = \frac{g}{2\pi}T_p^2 \tag{2.21}$$

where T_p is the peak time period.

The European Overtopping Manual was revised twice in 2007 and 2016, and recommended formulas were given in the form of Van der Meer and Janssen formulas (Equations 2.18–2.21).

In addition, some scholars have studied the average overtopping for the case of zero freeboard ($R_c=0$), including the formula given by Schüttrumpf and Oumeraci (2005):

$$\frac{q_w}{\sqrt{gH_{m0}^3}} = 0.0537\xi_{m-1,0} \quad \xi_{m-1,0} < 2.0 \tag{2.22}$$

$$\frac{q_w}{\sqrt{gH_{m0}^3}} = \left(0.0136 - \frac{0.226}{\xi_{m-1,0}^3}\right)\xi_{m-1,0} \tag{2.23}$$

where $\xi_{m-1,0}$ is the breaking parameter based on deepwater wavelength and mean energy period.

2.2.3 Distribution of individual overtopping volumes

Van der Meer and Janssen (1994) used the Weibull distribution with a shape factor of 0.75 and a scale factor a, which is dependent on the average overtopping discharge per wave and the overtopping probability to represent the distribution of water volume in individual waves. The probability distribution function is:

$$P_V = P(V_i \leq V) = 1 - \exp\left[\left(-\frac{V}{a}\right)^b\right] \tag{2.24}$$

with

$$a = 0.84\frac{T_m q_w}{P_{ow}} \tag{2.25}$$

where P_V is the probability of the overtopping volume per wave V_i being less than or similar to V, and $b=0.75$ is the shape factor.

Victor et al. (2012) conducted a detailed experimental study on the values of b to improve the knowledge on the probability distribution of individual wave overtopping volumes on steep ($0.36 < \cot\alpha < 1.69$, where α is the slope angle), low-crested ($0.10 < R_c/H_{m0} < 1.69$) smooth impermeable structures. The effect of non-Rayleigh-distributed incident waves, slope angle, relative freeboard, and wave steepness was discussed. The prediction formula is Equation (2.26):

$$b_V = \exp\left(-2.0\frac{R_c}{H_{m0}}\right) + (0.56 + 0.15\cot\alpha) \tag{2.26}$$

According to Pullen et al. (2007), b is likely to increase in shallow-water wave conditions. Nørgaard et al. (2014) modified Equation (2.26) to get a better prediction of b under shallow-water condition:

$$b = b_V C_{N1} \tag{2.27}$$

With

$$C_2 = \begin{cases} 1 & \text{for } H_{m0} / H_{1/10} \leq 0.848 \quad \text{or} \quad H_{m0} / h \leq 0.2 \\ -10.8 + 13.9 \cdot \dfrac{H_{m0}}{H_{1/10}} & \text{for } H_{m0} / H_{1/10} > 0.848 \quad \text{and} \quad H_{m0} / h > 0.2 \end{cases} \quad (2.28)$$

2.2.4 Probability of overtopping

The probability of overtopping is defined by the ratio of overtopping wave number N_{ow} and incoming wave number N_w as:

$$P_{ow} = \frac{N_{ow}}{N_w} \quad (2.29)$$

Besley (1999) gives the formula for prediction of the probability of overtopping in the design and assessment manual of seawalls:

$$P_{ow,B} = \begin{cases} 55.41 Q_*^{0.634} & 0 < Q_* < 0.008 \\ 2.502 Q_*^{0.199} & 0.008 \leq Q_* < 0.01 \\ 1 & Q_* \geq 0.01 \end{cases} \quad (2.30)$$

where Q_* is the dimensionless average overtopping discharge:

$$Q_* = \frac{q_w}{T_m g H_s} \quad (2.31)$$

Nørgaard et al. (2014) conducted a series of two-dimensional physical model tests on typical rubble mound breakwater geometries and provided a modification of the Besley (1999) formula, consulting the distribution of incident waves, to better predict the probability of wave overtopping under shallow-water condition:

$$P_{ow}^{\text{Shallow}} = P_{ow}^{\text{Besley}} \cdot C_1 \quad (2.32)$$

With

$$C_1 = \begin{cases} 1 & \text{for } H_{m0} / H_{1/10} \leq 0.848 \quad \text{or} \quad H_{m0} / h \leq 0.2 \\ -6.65 + 9.02 \cdot \dfrac{H_{m0}}{H_{1/10}} & \text{for } H_{m0} / H_{1/10} > 0.848 \quad \text{and} \quad H_{m0} / h > 0.2 \end{cases} \quad (2.33)$$

In the European Overtopping Manual another equation for the prediction of the probability of overtopping is:

$$P_{ow} = \exp\left[-\left(\sqrt{-\ln 0.02} \, \frac{R_c}{R_{u2\%}} \right)^2 \right] \quad (2.34)$$

where R_c is the freeboard, and $R_{u2\%}$ is the 2% run-up height.

2.3 Combined wave and surge overtopping

Relatively fewer studies have been conducted on wave overtopping that occurs when the water level is still higher than the levee crest. This case was referred to as "combined wave and surge overtopping" by Hughes and Nadal (2009). In this condition, the average overtopping discharge may be close to surge-only overflow discharge, but the peak instantaneous discharge can be several times the value of the surge-only overflow discharge. In recent years, noticing the catastrophic consequences of Katrina, some researchers have begun focusing on the study of combined wave and surge overtopping.

2.3.1 Combined wave and surge overtopping discharge

In Overtopping Manual (Pullen et al. 2007), the amount of water flowing to the landward side of the structure in combined wave and surge overtopping cases was divided into two parts: (1) a part attributed to surge-only overflow (q_s), and (2) a part attributed to wave-only overflow (q_w). The discharge of combined wave and surge overtopping, q_{ws}, can be calculated by Equation (2.35) as below:

$$q_{ws} = q_s + q_w \tag{2.35}$$

where q_s is calculated by Equation (2.2), and q_w is calculated by Equations (2.22) and (2.23).

However, because the nonlinear interaction between surge-only overflow and overtopping is strong, there must be some errors in Equation (2.35). In the aftermath of Hurricane Katrina, there is a lack of estimation method for combined wave and surge overtopping in the assessment and reinforcement of levees. In this book, Equation (2.35) is used as an emergency estimation method.

Reeve et al. (2008) built a numerical flume based on Reynolds-averaged Navier–Stokes equations to investigate the discharge characteristics of combined wave and surge overtopping of impermeable seawalls. The numerical flume was tested against published experimental observations, approximate analytical solutions, and empirical design formulas for surge-only overflow and wave-only overtopping. A sequence of combined wave and surge overtopping cases was simulated by the numerical flume. Based on the simulation results, equations for the dimensionless average discharge of combined wave and surge overtopping were provided for small negative freeboard ($0 > R \geq -0.8$):

For breaking waves ($\xi_p < 2$):

$$Q_R = \frac{q_{ws}}{\sqrt{gH_s^3}} \frac{\sqrt{\tan\alpha}}{\xi_p} = 0.051\exp\left(-1.98\frac{R_c}{H_s\xi_p}\right) \tag{2.36}$$

For nonbreaking waves ($\xi_p > 2$):

$$Q_R = \frac{q_{ws}}{\sqrt{gH_s^3}} = 0.233\exp\left(-1.29\frac{R_c}{H_s}\right) \tag{2.37}$$

where Q_R is the dimensionless combined wave and surge overtopping discharge. Based on the dimensionless overtopping discharge (Q_v) used by Van der Meer and Janssen, the average combined wave and surge overtopping discharge (q_{ws}) is used instead of the average wave-only overflow (q_w). Equations (2.36) and (2.37) are applicable at the ranges of $-0.8 < R_\xi < 0$. R_ξ and are defined as:

$$R_\xi = \frac{R_c}{H_s \xi_p}$$ (2.38)

Reeve's research (Reeve et al. 2008) was the first systematic study of combined wave and surge overtopping which proposed an empirical formula for estimating the discharge of combined wave and surge overtopping. However, according to the comparison with the flume test results (such as Hughes and Nadal's test, the experiment introduced in this book), the empirical formula provided overestimates the discharge of combined wave and surge overtopping. The reason for this could be that the Reeve's method is based on numerical flume, and the calibration and validation of numerical flume is based on surge-only overflow and wave-only overtopping. Thus, it cannot fully reflect the dynamic characteristics of combined wave and surge overtopping that can be observed in the field at different conditions.

Hughes and Nadal (2009) conducted a two-dimensional laboratory study on combined wave overtopping and storm surge overflow of a levee with a trapezoidal cross-section. The experiments were conducted in a 0.91-m-wide wave flume at a nominal prototype-to-model length scale of 25:1. Twenty-seven unique wave conditions for combined wave and surge overtopping were tested. A simple relation was sought between dimensionless overtopping discharge and relative freeboard, as given in Equation (2.39):

$$\frac{q_{ws}}{\sqrt{gH_{m0}^3}} = 0.034 + 0.53 \left(\frac{-R_c}{H_{m0}} \right)^{1.58}$$ (2.39)

The estimated results of Equation (2.39) are in good agreement with the experimental results under the condition of roller-compact concrete for landward-side slope in the 1:1 large flume test introduced in this book.

2.3.2 Distribution of individual overtopping volumes under negative freeboard

Based on the results of a series of 25:1 laboratory tests with negative freeboard, Hughes and Nadal (2009) developed new equations for the distribution of individual wave volumes of the water overtopping under negative freeboard, namely, combined wave and surge. Two-parameter Weibull distribution, as shown by Equation (2.24), was used to represent the distribution of individual wave volumes. The equations to predict the scale factor a (m³/m) and the shape factor b are given by the following equations:

$$a = 0.79 q_{ws} T_p$$ (2.40)

$$b = 15.7 \left(\frac{q_s}{g T_p H_{m0}} \right)^{0.35} - 2.3 \left(\frac{q_s}{\sqrt{gH_{m0}^3}} \right)^{0.79}$$ (2.41)

Because the probability of overtopping is 1, the overtopping discharge under negative freeboard is almost continuous except for some breaks between two large waves. Thus, the instantaneous overtopping discharge could also be represented by Weibull distribution. Hughes and Nadal (2009) provided Equations (2.42) and (2.43) for scale factor a_{in} (m^3/s/m) and the shape factor b_{in} based on the laboratory tests.

$$a_{in} = \frac{q_{ws}}{\Gamma\left(1+\dfrac{1}{b}\right)}$$ (2.42)

$$b_{in} = 8.10\left(\frac{q_s}{gT_pH_{m0}}\right)^{0.34}$$ (2.43)

The results of this research (Hughes and Nadal 2009) for the combined wave and surge overtopping are of great importance in the experimental study introduced in this book. In the experimental study introduced in this book, the estimation accuracy is further improved by further classifying the types of combined wave and surge overtopping based on the above formulas. See the related introduction in Chapter 5 for details.

2.3.3 Hydraulic parameters on the landward-side slope of levees

One of the main causes of failure of levees caused by combined wave and surge overtopping is the erosion and destruction of the landward-side slope of levees. Therefore, it is necessary to pay attention to the hydraulic parameters on the landward-side slope of levees during combined wave and surge overtopping. Based on the results of a series of 25:1 laboratory tests, Hughes and Nadal (2009) provided the calculation methods of mean flow thickness, RMS wave height, mean velocity, and velocity of the wavefront.

The mean flow velocity equation was:

$$v_m = 2.5\left(q_{ws}g\sin\beta\right)^{1/3}$$ (2.44)

where β is the landward-side slope angle.

The root-mean-square wave height was suggested to be calculated as:

$$\frac{H_{rms}}{d_m} = 3.43\exp\left(\frac{R_c}{H_{m0}}\right)$$ (2.45)

where d_m is the mean flow thickness on the levee slope.

A tentative estimation of wavefront velocity on the landward-side slope, v_w, was also given by Hughes and Nadal (2009):

$$v_w = 3.85\sqrt{gH_{rms}}$$ (2.46)

2.4 Turbulent overtopping

Overtopping volumes of water are highly turbulent and have a substantial amount of air entrainment (Hughes et al. 2012). Little is known regarding the more problematic and complex case of nonstationary high turbulent flow conditions. These turbulent flows significantly impact the design parameters for the various strengthening methods.

2.4.1 Turbulence measurement

Instruments commonly used to measure turbulence include microstructure profilers (e.g., Moum et al. 1995), hot-wire and hot-film anemometers (e.g., Chen and Chiew 2003), particle-image velocimeters (e.g., Meselhe et al. 2004), laser Doppler velocimeters (e.g., Nezu and Rodi 1986), and acoustic Doppler velocimeters (ADVs) (e.g., Garcia et al. 2005). Statistical descriptors of turbulence include mean velocity, turbulence intensities (standard deviation), turbulent kinetic energy (TKE), Reynolds stresses (a tensor formed from velocity fluctuation cross-correlations), spectra, integral time scale, eddy viscosity, and mixing length (Nystrom et al. 2003).

ADVs have been widely used in river engineering and turbulent open-channel flows to measure three-dimensional dynamic flows in physical models (Lohrmann et al. 1994; 1995; Voulgaris and Trowbridge 1998; Lane et al. 1998; Finelli et al. 1999; Synder and Castro 1999; Nikora and Goring 2000; Garcia et al. 2005; Carollo et al. 2005; Tricito and Hotchkiss 2005; Blanckaert and Lemmin 2006; Lacey and Roy 2008). The raw data collected by ADVs cannot be used directly because it contain Doppler noise and ambiguity spikes. The Doppler noise is associated with the measurement process itself, which is an inherent part of all Doppler-based volume backscatter systems (Lohrmann et al. 1995). Because the phase shifting between the outgoing and incoming pulse outside the range of −180 and +180°, aliasing of the Doppler signal causes a spike in the velocity record. Such a situation can occur when the flow velocity exceeds the preset velocity range or when there is a contamination from previous pulses reflected from the boundaries of complex geometries (e.g., cobbles on a stream) (Goring and Nikora 2002). Effective filtering methods have been proposed to restrain or remove noise and filter ADV velocity time series (e.g., Voulgaris and Trowbridge 1998; Blanckaert and Lemmin 2006; Garcia et al. 2005; Wahl 2000; Goring and Nikora 2002; Cea et al. 2007). An appropriate velocity range should be chosen to avoid velocity ambiguities using these filtering methods. The despiking method of Goring and Nikora (2002) had a good performance for detecting and replacing the spikes caused by the velocity ambiguities. This method was used in this study.

For turbulence measurements of flow in the presence of surface waves, surface waves can produce large biases in the estimation of turbulent normal and shear stress obtained from single-sensor measurements of velocity if there is even a small uncertainty in the orientation of either the velocity sensor or the principal axes of the wave-induced velocity field (Lumley and Terray 1983). Trowbridge (1998) suggested that the wave-induced bias can be diminished substantially by differencing measurements obtained from two velocity sensors separated by a distance larger than the correlation scale of the turbulence but small in comparison to the inverse wave number of the surface waves. Trowbridge and Elgar (2002) estimated scales of turbulence of near-shore tidal flow from a spatial array of velocity sensors.

2.4.2 Turbulent shear stress

Direct measurements of shear stress are rarely attempted and are different from indirect methods such as log profile method, Reynolds stress method, TKE k method, TKE w' method, and Nadal and Hughes method (Schlicting 1987; Dietrich and Whiting 1989; Kim et al. 2000; Babaeyan-Koopaei et al. 2002; Nadal and Hughes 2009).

The log profile estimates of stress are common in the field of coastal research and even used in the standard manner (Trowbridge 1998). The log profile method for the grass-covering bed surface is used to calculate local bed shear stress τ_0 from the logarithmic relation between the shear velocity and the variation of velocity with height (Stephan and Gutknecht 2002):

$$\frac{u}{u_*} = \frac{1}{\kappa} \ln\left(\frac{z - z_{p,m}}{z_{p,m}}\right) + 8.5 \tag{2.47}$$

where u is the velocity, u_* is the shear velocity $(=\sqrt{\tau_0/\rho})$, ρ is the water density, $\kappa\,(=0.41)$ is the von Kármán constant, z is the height above the bed, $z_{p,m}$ is the mean plant height, and 8.5 is the integration constant for a rough bed (Christensen 1985; Stephan and Gutknecht 2002). Although the log profile method is widely used, the velocity profile in complex flow may not be logarithmic (Babaeyan-Koopaei et al. 2002; Biron et al. 2004).

The Reynolds stress method is used to determine the local bed shear stress from Reynolds stress when the turbulence measurements are available:

$$\tau(z) = -\rho\overline{u'w'} \tag{2.48}$$

where $\tau(z)$ is the shear stress, u' and w' are the velocity fluctuations of the streamwise and vertical components, and the overbar denotes an average (Pope 2000; Babaeyan-Koopaei et al. 2002). With the development of ADV, it is possible to obtain detailed measurements of turbulent velocity fluctuations in the three components of velocity at high frequencies (McLelland and Nicholas 2000). When a turbulent profile is collected, bed shear stress is often obtained by extrapolating the vertical distribution of Reynolds shear stress to the bed (Nikora and Goring 2000; Song and Chiew 2001; Chen and Chiew 2003).

Linear relationships between turbulent energy and shear stress have been developed in a quasi-equilibrium turbulent energy model (e.g., Galperin et al. 1988). Soulsby and Dyer (1981) and analyzed turbulence data of tidal currents, showing that the average ratio of shear stress to TKE was constant. Soulsby (1983) extended this relationship and defined the TKE k method by including three components of velocity to calculate turbulence shear stress, as shown in Equation (2.49):

$$\tau(z) = C_1\rho k = C_1\rho\left[0.5\left(u'^2 + v'^2 + w'^2\right)\right] \tag{2.49}$$

where k is the TKE, $k = 0.5\left(u'^2 + v'^2 + w'^2\right)$, v' represents fluctuations of the cross-stream velocity, and C_1 is a proportionality constant (0.19). This method has been applied in shear stress analysis in estuarine, coastal flow, shelf, and oceanography (Stapleton and Huntley 1995; Williams et al. 1999; Kim et al. 2000; Huthnance et al. 2002).

The TKE w' method is a modification of the TKE k method (Kim et al. 2000). The TKE w' method only uses vertical velocity fluctuations because instrument noise errors associated with vertical velocity variances are smaller than noise errors for horizontal velocity fluctuations due to the use of ADVs to record the velocity data (Voulgaris and Trowbridge 1998; Biron et al. 2004). Thus, Equation (2.48) becomes:

$$\tau(z) = C_2\rho w'^2 \tag{2.50}$$

where C_2 is a proportionality constant (0.9) (Kim et al. 2000).

The Nadal and Hughes (2009) method is used to estimate the turbulent shear stress on the land-side levee facing nonuniform and unsteady overtopping flow by using the one-dimensional momentum equation (Equation 2.50):

$$\frac{\tau_0}{\rho g h} = \sin\theta - \frac{\partial h}{\partial s} - \frac{\partial}{\partial s}\left(\frac{u^2}{2g}\right) - \frac{1}{g}\frac{\partial u}{\partial t} \tag{2.51}$$

where g is the gravity, θ is the land-side levee slope angle, u is the instantaneous streamwise velocity, h is the water depth, s is the coordinate parallel to the land-side levee slope, and t is the time. This method needs time series of water depth and streamwise velocity to provide the time series of shear stress located between the two measured cross-sections.

Chapter 3

Three strengthening systems

3.1 Background

As engineers and authorities considered the modes of failure and the most cost-effective means to rebuild and restore the levee system, it became clear that armoring the land-side slopes of the levee system is vital to withstand the future hurricanes and the associated storm surges. In August 2007, the US Army Corps of Engineers conducted a "Levee Armoring Workshop" where design criteria were presented for the levees. These criteria are noted in Tables 3.1 and 3.2, which are established by the US Army Corps of Engineers and their consultants from the data gathered after Katrina to improve the system to withstand at least a 100-year storm (Villa 2007). The top three systems considered for levee armoring to meet these criteria include high-performance turf reinforcement mats (HPTRM), articulated concrete blocks (ACB), and roller-compacted concrete (RCC).

3.2 Three innovative levee-overtopping protection methods

The experimental study of combined wave and surge overtopping introduced in this book focuses on two aspects. One is the hydrodynamic characteristics of combined wave and surge overtopping process, and the other is the resistance of different slope protection methods to combined wave and surge overtopping. Three different kinds of slope protection methods were tested in this study: HPTRM, ACB, and RCC.

Table 3.1 Levee design criteria for the New Orleans hurricane protection system

Content	Criteria
Type of levee	Hurricane Protection Levee
Type of environment	Coastal/riverine
Storm surge	9 m (30 ft)
Design shear	720 Pa (15 lb/ft^2)
Design velocity	6.0 m/s (20 ft/s)
Levee design slope	3V:1H
Levee height	4.6–7.6 m (15–25 ft)
Levee width	60 m (200 ft)
Levee soil	Clay
Storm duration	2–6 hours
Soil loss (allowable)	25 mm (1 inch)

Table 3.2 Additional conditions for levee design

Content	Criteria
Armoring life	≥50 years, UV and Corrosion-Resistant Material
Vegetation cover	Bermuda, Rye, Pensacola Bahia Grasses
Environment	Levee slopes routinely submerged in fresh and saltwater
Maintenance	Levee boards mow grass approximately every 2 months
Construction	Levees to be constructed in lifts – Remove/reuse armoring to place next lift

3.2.1 High-performance turf reinforcement mats (HPTRM)

HPTRM is the most advanced flexible armoring technology available today for severe erosion challenges. This system combines extremely high tensile strength and superior interlock and reinforcement capacity. The HPTRMs use a unique three-dimensional matrix of nylon filaments with a high tenacity polyester geogrid reinforcement at low strains to lock soil in place and provide permanent reinforcement to prevent soil loss during storm events. This technology is specially designed to capture the moisture, soil, and water required for rapid growth of grass sods. The HPTRMs have extremely high tensile strengths, in the order of ten times to the traditional turf reinforcement mats (TRMs), as well as superior interlock and reinforcement capacity with either soil or root system. They are suited for tough erosion applications where high loading and/or high survivability conditions are required. Available recent data indicates that velocities up to 6.7 m/s can be tolerated using this system (Kelley and Thompson 2008). Figure 3.1 gives a structure illustration of HPTRM and the vegetated HPTRM system.

Prior to the advent of TRMs, vegetative linings were not considered for these highly erosive conditions where expected velocities would exceed 2.1 m/s (Chow 1959) or for shear stresses topping 177 Pa (Chen and Cotton 1988). However, modern TRMs have proven to substantially increase the erosion resistance of vegetation, enabling their use in areas where high velocities/shear stresses are prevalent (Hewlett et al. 1987; Northcutt and McFalls 1998). The growth and testing methods of TRMs have been discussed with a performance test by Lancaster (1996) and Lipscomb et al. (2003).

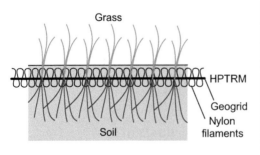 

Figure 3.1 Vegetated HPTRM system: structure illustration (a) and vegetated HPTRM system (b). (Adapted from Pan et al. (2015b). Reproduced with permission from Elsevier.)

Nelson (2005) conducted a study on the permissible shear stress of TRM based on a laboratory experiment in which TRMs were installed over a highly erodible sandy loam soil and subjected to 1-hour and 10-hour surge-only overflows generating escalating levels of shear stress. Excessive erosion was defined as the removal of an average of 12.7 mm of soil from beneath the mats and/or through the mat-reinforced vegetation in this study. This study determined that the permissible shear stress (900 N/m^2) for fully vegetated TRMs was much larger than the design values used in the previous studies. With the appearance of turf slope protection and erosion resistance, HPTRM has become a kind of slope armor with respect to both ecology and strength, which has been widely used in the world in recent years.

In designing any flexible HPTRM system, the following requirements should be considered: (1) strong, strain-insensitive, and flexible geosynthetics; (2) provision of significant reinforcement; (3) easy installation with minimum disturbance and without the necessity to grow new grass; (4) high cost-effectiveness and total concept; (5) invisible; (6) durable; and (7) no adverse environmental impacts (Akkerman et al. 2007). The anchored HPTRM system meets these requirements the best and provides additional positive features. One of the main advantages of the HPTRM is that it is a lightweight material and is particularly attractive when conditions dictate that a minimum load be placed on the levee and/or foundation. Other advantages include maintenance access, applicability to steep slopes, and can be used and installed easily in both arid and semi-arid environments.

Earth percussion anchors are made of corrosion aluminum alloy and gravity die-cast and are heat-treated to provide a considerable increase in mechanical strength and durability. The anchors are connected to a galvanized threaded rod or stainless tendon to enhance corrosion resistance, particularly at the soil/air interface. As load is exerted on the soil by the anchor, a body of soil above the anchor is compressed and provides resistance to any further anchor movement, and thus, permanently securing the mat to the ground.

The disadvantages of vegetated HPTRM are that it takes several months (or years) for the root system to develop and is limited (unless more flood-resistant grass and/or additional anchorage was provided) to flow durations less than 2 days. In addition, reinforced grass can handle flow velocities of up to 3.8 m/s. Although reinforced grass is generally more economical than conventional engineering materials in capital cost, it can be more expensive to maintain. In summary, HPTRM provides a cost-effective and lightweight solution for relatively high flow conditions.

In this study, the vegetation is warm-environment Bermuda grass. The HPTRM product selected for this study was Enkamat R30 from Colbond Inc. As indicated in Figure 3.2, Enkamat R30 is a three-dimensional TRM joined at the intersections of randomly oriented nylon filaments with high tenacity polyester geogrid reinforcement at low strains. The HPTRM is a high-performance geosynthetic composite with 95% open space. As the roots grow through the open space of HPTRM, they become entwined within the TRM. The interlocking between roots and HPTRM can enhance the resistance of the roots against hydraulic life and shear forces created by high water flow hydraulic erosion. The specific gravity of nylon (in the HPTRM) of more than 1 ensures that the HPTRM will not float under any hydraulic condition. The geogrid reinforcement in the HPTRM can help soil stabilization mechanically by taking over when extreme conditions exist. U-shaped pins are used as anchors to lock the HPTRM into the soil.

Figure 3.2 Details of the three-dimensional structure of HPTRM. Polyamide filaments thermally fused at the intersections, and polyester fibers interwoven as a geogrid interlock. (Adapted from Pan et al. (2015b). Reproduced with permission from Elsevier.)

Table 3.3 lists the mechanical properties of the HPTRM provided by the manufacturer. The performance properties of the HPTRM are also shown in Table 3.4.

3.2.2 Articulated Concrete Block (ACB) system

In recent years, ACBs have provided innovative and cost-effective solutions for embankment dam protection. A cost-effective method for protecting the land-side is to allow overtopping and use concrete blocks (hard-armor products) designed for hydraulic stability to protect the downstream slope under high-stress applications (high flow). An ACB system is a matrix of machine-compressed individual concrete blocks assembled to form a large mat. Blocks are 10–23 cm thick and 0.093–0.186 m^2 in plan with openings penetrating thought the entire block. Blocks are usually designed to be intermeshing or interlocking, and many units are patented. The matrix is connected by a series of cables which pass longitudinally through performed ducts in each block, making them easy to install over site-specific filter fabric on a prepared surface. Blocks may be solid or have open cells to permit uplift pressure relief and vegetation growth (Fuller 1992).

Some positive features of concrete block systems include: (1) the ability to sustain relativity high flow velocities (in excess of 7.6 m/s) immediately following installation; (2) ability of blocks with open cells to release excess hydrostatic pressures; and (3) ability to accommodate small subgrade changes caused by settlement, frost heave, and surface slumping (Clopper 1991; Koutsourals 1994). In addition, the hydraulic stability of the system is independent of flow duration. On the other hand, the disadvantages of concrete block systems are that the interlocking feature between units must be maintained. Routine maintenance is also required to prevent bushes from growing through the openings. An underlying geotextile and anchor system may be required depending on subsoil conditions (e.g., Powledge and Pradivets 1992).

The ACB system used in this study included individual ACB, cable, geogrid, gravel, and geotextile. There are several ACB vendors in the market. Armorflex 70-T

Table 3.3 Mechanical properties of the HPTRM from manufacturer

Mechanical properties	Test method	Units	Typical roll value
Ultimate tensile strength	ASTM D6818	kN/m	30
Tensile modulus at 2% strain	ASTM D6818	kN/m	265
Thickness	ASTM D6525	mm	19
Mass/unit area (TRM+Grid)	ASTM D6566	g/m^2	450
UV stability (1,500 hours)	ASTM D4355	%	80

Table 3.4 Performance properties of the HPTRM from manufacturer

Performance properties	Test method	Units	Typical roll value
Permissible velocity			
30 minute, unvegetated	Flume test	m/s	4.9
60 minute, vegetated	Flume test	m/s	6.1
50 hour, vegetated	Flume test	m/s	4.2
Permissible shear stress			
30 minute, unvegetated	Flume test	kN/m^2	0.28
60 minute, vegetated	Flume test	kN/m^2	0.81
50 hour, vegetated	Flume test	kN/m^2	0.38
Manning's n Range	Flume test		0.025–0.045

manufactured by Contech Inc. was used as a tapered ACB. ACB 70-T concrete block is an open cell for high-velocity application block. The individual cell has a typical dimensional of 44.2 cm $(L) \times 39.4$ cm $(W) \times 21.6$ cm (H), 0.164 m^2 gross area, and 20% open area. The tapered block design of a 12.7-mm taper allows for water to cascade over the top of the blocks in a shingle-like fashion limiting the projection forces on the block during an overtopping event and eliminates destabilizing impact flow forces to provide higher safety. The tapered block has a downstream thickness of 12.7 mm taller than the upstream end. Figure 3.3 shows the ACB 70-T.

The typical weight of one block is 54.5–62.7 kg and the unit weight is 333–382 kg/m^2. The 70-T is the heaviest block in the high-velocity application block classes. The unit weight of one block is 2,000–2,400 kg/m^3. The compressive strength of the block is 24.1–27.6 MPa, and the maximum water absorption is 192 kg/m^3 according to ASTM C140.

The system is a flexible, interlocking matrix of concrete blocks of uniform size, shape, and weight connected by a series of cables which pass longitudinally through preformed ducts in each block. The blocks are interconnected by flexible cables, providing articulation between adjacent blocks. Block walls are designed with beveled sidewalls to allow for flexibility in all directions. ACB is installed over geogrid, gravel, and geotextile on a prepared soil surface. The permeability of the ACB system including cell opening and periphery of blocks relieves hydrostatic pressure on the soil surface. Armoring the embankment with ACB helps to prevent erosion caused by steep-gradient and high-velocity flow.

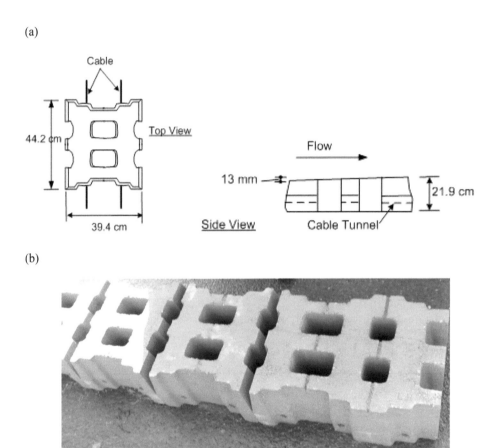

Figure 3.3 Articulated concrete block: (a) dimension of ACB block, and (b) photo of ACB block.

The Manning roughness coefficient, n, has a value of 0.026–0.034, depending on the configurations and type of the block used. According to the manufacturer data, the maximum flow velocity of ACB blocks that can resist is 5.9 m/s, and the maximum shear stress is 570 N/m^2. The ACB system combines the favorable aspects of lightweight geotextile and geogrid, such as porosity, flexibility, vegetation encouragement, and habitat enhancement with nonerodible, self-weight, and high shear force resistance of a rigid lining. The ACB system has proven to be an aesthetic and functional alternative to dumped stone riprap, gabions, structural concrete, and other heavy-duty, durable erosion protection systems. The ACB system is easy to install, and therefore, can dramatically reduce overall project costs. Specifically, compared to other systems, the life-cycle costs can be reduced as the ACB system is permanent, and significant savings on subsequent maintenance expenses can be achieved. ACB can be backfilled with topsoil and seed for vegetation and is a cost-effective, aesthetically pleasing, engineered solution for levee-overtopping protection.

3.2.3 Roller Compaction Concrete (RCC)

RCC is formed by a mixture of controlled-gradation aggregate, Portland cement, and possible pozzolans (fly ash), mixed with water and then compacted by a roller (e.g., Hansen and Reinhardt 1991; Choi and Hansen 2005). RCC has higher strength properties and abrasive resistance than cement-stabilized soils. The main difference between RCC and conventional concrete is that RCC has an aggregate gradation and paste content suitable for compaction by a vibratory roller (McDonald and Curtis 1997; Choi and Hansen 2005). In addition, conventional concrete has formed or screened surface imperfections that cause cavitation erosion at high (~12 m/s) velocities (ASCE 1994). The major advantages of RCC include reduced cost and speed of construction. RCC has been used in dam construction and/or modification and has potential application for use in protection against levee-overtopping (ASCE 1994; Hansen 2002; Choi and Hansen 2005).

RCC may be placed and compacted in stairstep horizontal layers or by plating a single layer placed parallel to the slope. While the plating method uses less material and is more economical, it is difficult to install on steep (20%or greater) slopes, and the smooth face results in greater wave run-up (Bingham et al. 1992). When the plating method is used, it should be placed on the river-side to prevent head cutting at the land-side to prevent slippage down the slope. For critical locations, a geotextile and/or gravel drainage layers may be required to reduce uplift pressures beneath the RCC (Perry 1998). Figure 3.4 gives an example of the RCC dam compaction construction (USDOI 2017).

Erosion of RCC used in the overflow protection of levees can occur due to hydraulic shear stress exerted by the flowing water; abrasive action from sand, gravel, or other waterborne debris; and cavitation from surface imperfections at flow velocities as low as 12.2 m/s (ASCE 1994). The method of construction employed for RCC may result in surface imperfections. Therefore, cavitation erosion would be expected if flow velocities exceed 12.2 m/s.

The upper limit of flow velocity (below 12.2 m/s where cavitation erosion occurs) above which appreciable erosion occurs for RCC has not been established. Erosion experiments conducted on RCC in the laboratory indicate that for clear water flow (no abrasive action), if any, erosion occurred for flow velocities of up to 10.7 m/s (limitations imposed by equipment) and for flow durations of up to 20 hours (Saucier 1984).

Figure 3.4 Compression works of RCC dams: Cold Springs dams (a) and Pueblo dams (b). (After USDOI (2017).)

In addition, RCC with increased strength and larger size aggregates could be used for coarse sediment present in the overtopping flow.

The density of RCC is ranges 2,320–2,430 kg/m^3. The compressive strength of RCC ranges 6,890–27,580 kPa. Split tensile strength of RCC is 5%–10% of compressive strength cured in the same period. The primary property used in RCC design is compressive strength. The required compressive strength is 20,684 kPa at 3–5 days of curing for this study. The mixing component of RCC included sand (1,050 kg/m^3), ¾″ gravel (1,030 kg/m^3), pozzolan (270 kg/m^3), cement (60 kg/m^3), and water (50 kg/m^3).

3.2.4 Environmental impact of three levee-strengthening systems

The environmental impact of utilizing the "soft-armor" alternative of the HPTRM has a tremendous "green" factor compared to the "hard-armor" alternatives of RCC and ACB systems. A fully vegetated HPTRM on the face of the levee is considerably eco-friendlier than the hard-armor alternatives by providing localized cooling and a more accommodating habitat for wildlife. The ACB may be less onerous than the RCC in this regard assuming even partial vegetation establishment, which is common in these applications. However, the obvious thermal impact of the paved concrete surface is very significant for contributing to localized warming, degrading the otherwise healthy wildlife habitat. While the main purpose of the levee is structural and to prevent flooding and protect against erosion-related soil loss, positive environmental impact and system sustainability are certainly measures of stewardship that should weigh heavily in any engineering calculation and design. The estimated total CO_2 footprint for the HPTRM levee-strengthening system is the lowest among the three levee-strengthening systems discussed here (Goodrum 2011).

3.3 Material properties of the three strengthening systems

Laboratory testing was conducted to determine the physical and mechanical properties of the three levee-strengthening materials mentioned above.

3.3.1 Testing methods and properties results of HPTRM

The fibers that form the square pattern in HPTRM material were not symmetric with respect to the longitudinal and the axial directions. The longitudinal fibers consisted of four strands placed sideways, whereas the axial fibers had four strands that were interwoven (Figure 3.5). The structure of the interwoven material for an axial fiber is shown in detail in the cross-sectional scanning electron micrograph (SEM) of a metallographically prepared cross-section in Figure 3.6. An SEM image of the polymer fibers themselves is shown in Figure 3.6a. The smaller fibers in Figure 3.6b are responsible for the strength properties of HPTRM.

Thickness, mass per unit area, light penetration, ultimate tensile strength, tensile strength at 2% strain, tensile elongation, resiliency, flexibility, stiffness, and viscosity of selected HPTRMs were measured using the relevant ASTM standards and corresponding testing protocols. A mass per unit area measurement was performed on the HPTRM material by cutting 14 pieces of dimensions 15.2×15.2 cm. The mass of the pieces was measured using a very sensitive scale. The tensile strength testing of the

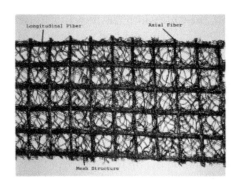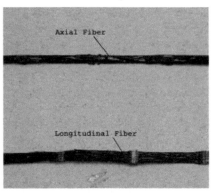

Figure 3.5 Details of the Enkamat HPTRM (a). The axial fiber and the longitudinal fibers are made of polymeric materials with the difference that the axial fibers are interwoven (b).

(a)

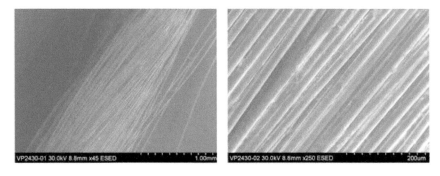

(b)

Figure 3.6 SEM images of the constituents of the HPTRM. (a) Images of the metallographically prepared cross-section of an axial fiber. (b) SEM images of the smaller polymeric fibers form the core of the individual strands.

HPTRM was conducted using the wide -width tensile testing method (ASTM D 4595) and the roller testing grips method (ASTM D 6637). As there was slippage from the grips or failure close to the gripping section, the wide width tensile testing method was inadequate for HPTRM testing. The roller testing grips method was then used for the HPTRM material. The specimens were 20.3 cm wide and 182.9 cm long with a cross-head displacement rate of 10% per minute. The strain calculation for roller clamps was based on a gage length of 10.2 cm and was measured. A slack correction of 80 N was used. The flexural rigidity (stiffness) of the HPTRM was measured according to ASTM D 6575. The 40.6-cm long specimens with a width of 10.2 cm were slid parallel to their length so that their ends were projected from the edge of a horizontal surface. The overhang length was measured when the tip of the test specimen was depressed due to gravitational forces such that the line joining the tip to the edge of the platform made a 41.5° angle from the horizontal plane. The bending length was half of the overhang length. The resiliency and apparent viscosity under ambient conditions were measured using ASTM D6524.

The average mass per unit area value of the HPTRM was determined to be 665.64 ± 31.63 g/m^2. The thickness of the HPTRM material was measured as 1.66 ± 0.07 cm, and the mass per unit area was measured as 665.6 g/m^2.

The tensile strength testing of the HPTRM was conducted with roller testing grips method (ASTM D 6637), as shown in Figure 3.7. Using the roller testing grips method, the load-strain curves for the Enkamat material in the machine and cross-machine direction are shown in Figure 3.8. A summary of the test results for the 2% tensile modulus (kN/m), ultimate tensile strength (kN/m), and elongation to failure (%) along with the average and standard deviation values are listed in Table 3.5. The results shown in this table are for the machine and cross-machine directions. All these results were obtained for 20.3-cm wide and 182.9-cm long specimens with a crosshead displacement rate of 10% per minute. The strain calculation for roller clamps was based on a gage length of 10.2 cm. A slack correction of 80 N was used.

The side rub ruptured first

Figure 3.7 Illustration of failure initiation in the outer rib for a multirib geotextile.

(a)   (b)

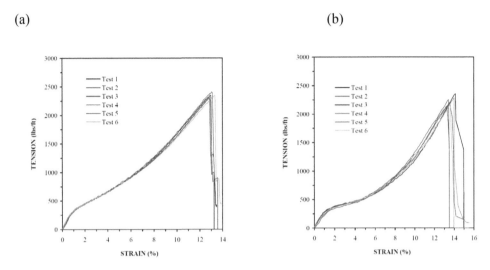

Figure 3.8 Stress–strain curves generated using roller grips for HPTRM material in the (a) machine direction and (b) cross-machine direction. The strains were calculated with a gage length of 10.2 cm.

Table 3.5 Tensile test results in the machine direction and cross-machine direction

Test number	Tensile modulus at 2% strain (kN/m)	Ultimate tensile strength (kN/m)	Elongation to failure (%)
In the machine direction			
1	319.7	34.3	13.0
2	329.2	35.2	13.1
3	327.8	33.7	12.9
4	335.1	33.9	12.9
5	319.7	33.6	12.8
6	327.8	34.3	13.3
Mean	326.6	34.2	13.0
Standard deviation	5.9	0.6	0.2
In the cross-machine direction			
1	279.6	34.7	14.2
2	272.3	33.1	13.5
3	251.1	31.3	13.3
4	267.2	33.4	14.0
5	264.3	32.2	13.5
6	264.3	31.9	13.6
Mean	266.5	32.8	13.7
Standard deviation	9.5	1.2	0.3

The results in Table 3.5 indicate that the HPTRM material had a slightly higher mean tensile modulus and mean tensile strength in the machine direction compared to the cross-machine direction. The elongation to failure, however, was marginally higher in the cross-machine direction. The t-tests conducted on both the ultimate tensile strength and elongation to failure indicated that the differences in the values between the machine and cross-machine directions were significant at 95% confidence level.

The HPTRM material had an average flexural rigidity of 139,800 g±4.1 mg-cm. The material had an average thickness of 1.66±0.07 cm and an average resiliency of 67.9%±2.6%. The average apparent viscosity under ambient conditions of the material was $1.68±0.34×10^8$ Pa-s. The light penetration for the HPTRM was determined to be 42.37±0.71 (%).

A summary of thickness and resiliency measurements on HPTRM specimens is presented in Table 3.6. Figure 3.9 illustrates the loading and unloading dwell cycles applied to a TRM specimen for thickness and resiliency tests.

A summary of bending length measurements on the HPTRM specimens is included in Table 3.7. Figure 3.10 illustrates the time-dependent changes in extension and load during compression.

A summary of the apparent viscosity of the HPTRM specimens is included in Table 3.8. Figure 3.11 illustrates the compressive stress as a function of strain rate.

Table 3.9 summarizes the properties of the HPTRM used in this study.

Table 3.6 Summary of thickness and resiliency measurements on five HPTRM specimens

Coupon ID	Initial thickness (T_i-cm)	Final thickness (T_f-cm)	Resiliency (%)
O2	1.69	1.16	68.2
O3	1.59	1.14	71.5
O4	1.74	1.13	64.7
O5	1.71	1.13	66.4
O6	1.59	1.09	68.5
Mean	1.66	1.13	67.9
Standard deviation	0.07	0.03	2.6

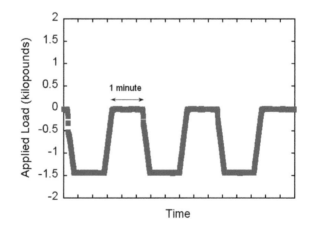

Figure 3.9 The loading and unloading dwell cycles applied to an HPTRM specimen for thickness and resiliency tests.

Table 3.7 Summary of bending length measurements on five HPTRM specimens

Coupon ID	Average bending length (c in cm)	c^3 (cm^3)	Flexural rigidity G (mg-cm)
5	12.98	2,186.88	145,558.4
6	12.71	2,053.23	136,662.7
7	12.81	2,102.07	139,913.8
8	12.72	2,058.08	136,985.5
Mean	12.81	2,100.1	139,780.1
Standard deviation	0.13	61.90	4,120.5

(a)

(b)

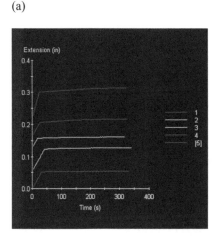
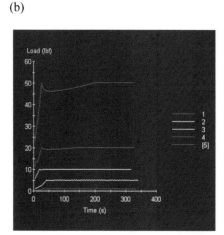

Figure 3.10 Time-dependent changes in (a) extension and (b) load during compression testing in HPTRM specimens. Conditions 1, 2, 3, 4, and 5 correspond to constant loads of 4 N, 22 N, 45 N, 89 N, and 223 N, respectively.

Table 3.8 Summary of the apparent viscosity of HPTRM specimens

Coupon ID	Apparent viscosity (Pa-s)
01R	1.06×10^8
02R	2.06×10^8
03R	1.60×10^8
04R	1.86×10^8
05R	1.81×10^8
Mean	1.68×10^8
Standard deviation	0.34×10^8

3.3.2 Testing methods and properties results of ACB

Two 70-T ACBs were tested to determine the physical and mechanical properties of ACB. These properties include absorption, density, and compression tests that were conducted according to ASTM C140 and ASTM C6684. Ten rectangular parallelepiped

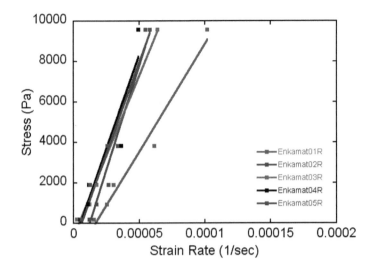

Figure 3.11 Compressive stress versus strain rate plots for five HPTRM specimens. The slope of these curves gives the viscous flow properties of this material.

Table 3.9 Summary of HPTRM properties and the testing methods

Physical and mechanical property	Testing method	Test results
Mass per unit area	ASTM D6566	665.6 ± 31.6
Tensile strength	ASTM D6637	34.2 ± 0.6
Tensile strength at 2% strain	ASTM D6637	326.6 ± 5.9
Tensile elongation (%)	ASTM D6637	13.0 ± 0.2
Resiliency (%)	ASTM D6524	67.9 ± 2.6

specimens were sawed and prepared to approximate dimensions of 15.2 cm length×3.3 cm thickness×7.6 cm height. The height dimension of the test specimens was in the same direction as the ACB unit's height dimension. The compression testing specimens were capped prior to testing according to ASTM C1552. Additional specimens were core-drilled from the second ACB unit to determine the shear strength by direct shear strength testing that was conducted according to ASTM D5607. Five specimens were extracted from both horizontal and vertical directions within the block. The specimens were approximately 4.39 cm in diameter and 8.89 cm in height. The normal load values during testing varied between 445 and 890 N.

The dimensions and weights, as well as density and water absorption test results for the ACB, are summarized in Tables 3.10 and 3.11, respectively. The average value of water absorption was 131.4 kg/m^3, which was below the maximum water absorption allowed. The average density was 2,128.3 kg/m^3, which was above the minimum density requirements for this type of ACB. The ACB unit passed the relevant requirements for water absorption and density values.

Table 3.10 Dimensions and weights of the ACB specimens for absorption and density calculations

Sample ID	Average length (cm)	Average height (cm)	Average thickness (cm)	Saturated weight (kg)	Immersed weight (kg)	Dry weight (kg)
A1	15.291	7.686	3.970	1.023	0.5679	0.9606
A2	15.644	7.770	3.729	1.0198	0.5714	0.9626
A3	15.532	7.658	4.107	1.0714	0.6015	1.0139
A4	15.453	7.722	3.945	1.0310	0.5680	0.9637
A5	15.481	7.757	3.780	1.0169	0.5715	0.9627

Table 3.11 Water absorption and density values for the five ACB specimens

Sample ID	Absorption (%)	Absorption (kg/m^3)	Density (kg/m^3)	ASTM C 6684-04 specification			
				Maximum water absorption (kg/m^3)		Minimum density (kg/m^3)	
				Average of 3 units	Individual unit	Average of 3 units	Individual unit
A1	6.5	137.75	2,107.2				
A2	5.9	128.14	2,144.0				
A3	5.7	121.74	2,153.6	...	187.41	...	2125
A4	7.0	145.76	2,078.4				
A5	5.6	121.74	2,158.4				
Average	...	131.35	2,128.3	145.76	...	2,080	...

During the compression testing, the loading rate was adjusted based on the ASTM requirements so that the total test time was between 3 and 5 minutes. Figure 3.12 shows the capped ACB specimen before and after the compression test. A summary of the compression test results is provided in Table 3.12. The average compression strength of the specimens was approximately 48,263.30 kPa, which was much higher than the listed minimum requirements based on the ASTM standard.

As shown in Table 3.13, the average shear strength was determined to be a function of the applied normal load and orientation. The shear strength values of specimens that were parallel to the vertical direction of the ACB were determined to be 1,142 and 2,066 kPa for normal loads of 445 and 890 N, respectively. The shear strength values of specimens that were parallel to the horizontal direction of the ACB were determined to be 1,823 and 2,671 kPa for normal loads of 445 and 890 N, respectively.

Table 3.14 shows the properties of the ACB used in this study.

3.3.3 Testing methods and properties results of RCC

The approximate dimensions of the RCC slab were 0.91 m $(L) \times 0.91$ m $(W) \times 0.37$ m (H). A construction lift joint was observed at approximately 13 cm from the top surface of the slab. Specimens were cored from the slab with approximate dimensions of 9.4 cm diameter and 19.6 cm length, as shown in Figure 3.13.

Figure 3.12 Specimen compression test after failure.

Table 3.12 Summary of the compression tests performed on the ACB specimens

Sample ID	Area (cm^2)	Ultimate load (kN)	Compression strength (kPa)	ASTM C 6684-04 specification	
				Minimum compressive strength (kPa)	
				Average of 3 units	Individual unit
CI	59.42	305.72	51,435.21		
C2	59.68	288.64	48,401.50		
C3	60.90	283.04	46,470.95	...	24,131.80
C4	61.42	327.43	53,296.80		
C5	59.87	265.11	44,264.62		
Average	...	...	48,263.60	27,579.20	...

Splitting tensile strength testing was performed according to ASTM C496 on cored specimens that included the construction lift joint. The lift joint was placed in the center of the split tensile specimen. Compression testing was performed according to ASTM C42 on cored specimens that did not include the construction lift joint. The density of the specimens was measured using a mass per unit volume calculation following ASTM C1176.

Figure 3.14 shows the image of a split tensile test specimen before and after testing. A summary of the split tensile strength test results is presented in Table 3.15. The average split tensile strength of five RCC specimens was 2,608 kPa.

Table 3.13 Summary of the direct shear strength tests performed on the cylindrical ACB specimens

Specimen ID	Diameter (cm)	Area (cm²)	Normal load (N)	Shear load (kN)	Shear strength (kPa)	Average shear strength (kPa)
1-Vertical	4.39	15.16	445	1.874	1,235.55	
2-Vertical	4.42	15.35	445	2.033	1,323.80	1,141.78
3-Vertical	4.39	15.16	445	1.314	866.68	
4-Vertical	4.42	15.35	890	3.444	2,242.88	
5-Vertical	4.42	15.35	890	2.906	1,891.24	2,065.68
1-Horizontal	4.39	15.16	445	2.932	1,933.99	
2-Horizontal	4.39	15.16	445	2.564	1,690.61	1,822.99
3-Horizontal	4.39	15.16	445	2.795	1,843.67	
4-Horizontal	4.39	15.16	890	4.655	3,070.25	
5-Horizontal	4.39	15.16	890	3.444	2,271.15	2,671.05

Table 3.14 ACB properties and the testing methods

Physical and mechanical property	Testing method	Test results
Shear strength (kPa)	ASTM D5607	1.5 ± 0.6
Compressive strength (kPa)	ASTM C42	48.8 ± 3.7
Density (kg/m³)	ASTM C6684	20.9 ± 0.3

Figure 3.13 Image of RCC cored specimens. (Adapted from Li et al. (2012). Reproduced with permission from Elsevier.)

Compression testing was performed according to ASTM C42 on cored specimens that did not include the construction lift joint (Figure 3.15). The length-to-diameter ratio was, therefore, such that a correction factor was applied to some specimens. The correction factor along with the compression strength results are summarized in Table 3.16. The average compression strength for five specimens was 22,656 kPa. The compression strength of the RCC specimens was less than half of the compression strength of ACB specimens. The aggregate structure and manufacturing process are responsible for the difference in the compression strength of these two kinds of materials.

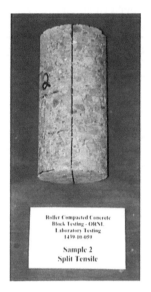 

Figure 3.14 Image of an RCC split tensile test before and after testing. The construction lift joint in the center of the specimens can be noticed.

Table 3.15 Summary of split tensile test results

Specimen number	Diameter (cm)	Length (cm)	Density (kg/m³)	Maximum load (kN)	Split tensile strength (kPa)
1	9.58	19.57	2,199.27	89.63	3,033.71
2	9.58	19.81	2,192.86	82.02	2,757.92
3	9.58	19.30	2,165.63	70.10	2,413.18
4	9.58	19.30	2,196.07	65.52	2,240.81
5	9.40	19.30	2,170.44	73.35	2,585.55

Table 3.16 Summary of compression test results

Test number	Diameter (cm)	Length before cap (cm)	Length after cap (cm)	Density (kg/m³)	L/D ratio	Maximum load (kN)	Correction factor	Compressive strength (kPa)
1	9.40	16.76	17.27	2,109.57	1.84	141.10	1.00	19,788
2	9.53	16.76	17.27	2,120.78	1.81	143.54	1.00	20,064
3	9.55	15.49	16.00	2,178.45	1.68	176.42	0.97	23,718
4	9.58	18.29	18.54	2,114.38	1.94	150.17	1.00	21,650
5	9.40	12.45	12.70	2,205.68	1.35	212.71	0.94	28,062

The density of the specimens was measured based on a mass per unit volume calculation. The average density of ten specimens was $2,165.63 \pm 36.84$ kg/m³.

In summary, Table 3.17 shows the properties of the RCC used in this study.

Figure 3.15 Image of an RCC specimen compression test before and after testing.

Table 3.17 RCC properties and the testing methods

Physical and mechanical property	Testing method	Test results
Splitting tensile strength (kPa)	ASTM C496	2.6 ± 0.3
Compressive strength (kPa)	ASTM C42	22.7 ± 3.4
Density (kg/m³)	ASTM C6684	21.2 ± 0.4

Full-scale physical model testing of levee overtopping

This chapter describes a physical model study of a full-scale (1:1) model of levee overtopping by combined wave and surge overtopping. The construction, instrumentation, testing procedures, and initial data analysis of the test sections are included in this chapter. The scale effect and measurement effect of the model are also discussed briefly in this chapter.

4.1 Full-scale test model setup

Full-scale flume tests were designed following two main research objectives: determining the hydraulic and erosion characteristics of levees under combined wave and surge overtopping.

4.1.1 Test facility

The full-scale physical testing was conducted in the Large Wave Flume (LWF) of the O.H. Hinsdale Wave Research Laboratory (HWRL) at Oregon State University. The flume is the largest of its kind in North America. It measures 104-m long by 3.7-m wide by 4.6-m wide, as shown in Figure 4.1. The wavemaker is a large-stroke, piston-type wavemaker capable of generating periodic or random waves to simulate the wave spectra associated with large storms. The tests assume a two-dimensional flow (waves are normal to the specimen). The LWF is capable of generating a 1.7-m high wave at a 3.5-second frequency, in 3.5 m water depth. Because of its size and ability to operate in high Reynolds regimes, the flume is ideally suited for levee testing under true full-scale overtopping conditions.

4.1.2 Levee embankment setup

The levee embankment was set up as a full-scale (1:1) model, as shown in Figure 4.2. The embankment consists of a crest, water-side slope, and landward-side slope, with dimensions of 26.13 m long×3.25 m high×3.66 m wide. The water-side slope has a 1V:4.25H slope, and the landward-side slope has a 1V:3H slope. The embankment was located 44.28 m downstream from the wavemaker gate to the toe of the water-side slope, as shown in Figure 4.3. The levee model was built with a sand core and a concrete cap with concrete of nominal 0.152 m thickness, with a center section of the crest and of the landward-side slope to be used as a test section for evaluating the three

Figure 4.1 Large Wave Flume at O.H. Hinsdale Wave Research Laboratory.
Source: https://wave.oregonstate.edu/.

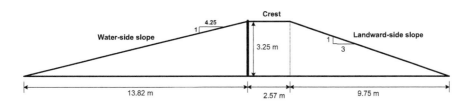

Figure 4.2 Profile of full-scale levee embankment model. (Adapted from Li et al. (2012). Reproduced with permission from Elsevier.)

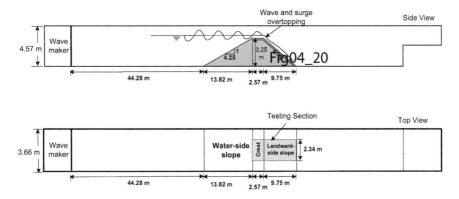

Figure 4.3 Location of levee embankment in the large wave flume. (Adapted from Li et al. (2012). Reproduced with permission from Elsevier.)

levee-strengthening systems. The test section overall dimensions were 12.32 m long, 2.34 m wide, and 0.76 m deep, which was used to install the different protection layers (Figure 4.4). Lean clay soil was placed in the test section and compacted at the maximum dry unit weight of 16.1 kN/m^3 and optimum water content of 19%.

4.1.3 Wave generator

To produce a true full-scale testing condition, a wavemaker was used. The wavemaker is a single-channel, piston-type hydraulically driven system. It consists of a vertical wall, referred to as the wavemaker piston, which is suspended from a steel support structure. The piston and its support structure travel back and forth on linear bearings that are bolted to the walls of the flume. The piston is driven by a set of counterbalanced hydraulic actuators that can exert up to 222.5 kN of force with a maximum hydraulic flow rate of 0.04 m^3/s. The piston can move at up to 4 m/s and has a 4 m total stroke, which is defined as the distance between its most offshore and most onshore positions.

As shown in Figure 4.5, the single-piston-type wavemaker can generate a maximum wave height of 1.7 m at $T=2.5$–5.0 s. It can generate regular waves, irregular waves with standard spectral shapes, solitary waves, and user-defined waves. Active absorption is available for regular and irregular waves. The active wave absorption was implemented during the testing.

Irregular waves were used to simulate the random wave condition in wave overtopping. Figure 4.6 presents an example of the generated wave height as a function of time. The significant wave height (H_{m0}) and peak wave period (T_p) were provided to the wavemaker control panel by the user. The generated random wave was a unidirectional TMA-type shallow water wave spectrum with a standard shape factor γ of 3.3. The manufacturer

Figure 4.4 Concrete cap of levee embankment in the crest and landward-side slope (a) and soil compaction for the 45-cm deep clay layer (b).

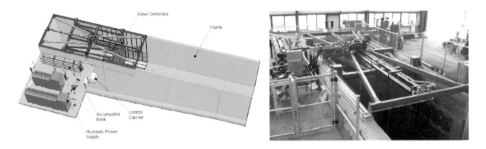

Figure 4.5 Scheme of large-stroke, piston-type wavemaker (a) and image of the wave-maker (b). (image courtesy of MTS Corporation.)

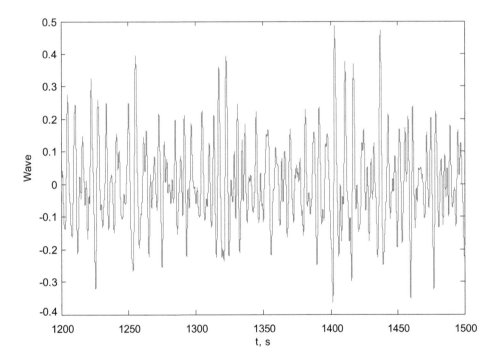

Figure 4.6 Illustration of generated random wave with user-specified significant wave height and peak wave period.

(MTS Systems Corporation) calibrates the wavemaker annually. The generated spectrum was then used with randomly phased waveforms to generate a unidirectional, multispectral free surface time series. The zeroth-moment spectral wave height (H_{m0}) was used. The wavemaker accepts a single channel of free surface time-series input, which is then internally converted using linear wave theory within the wavemaker control software to produce board motion. Prior to the running of any waves, a given free surface time-series input was previewed to ensure that kinematic limits would not be exceeded.

The data acquisition system collected all board motion, start/stop signals, water depth, and board free surface elevation during operations.

The wavemaker accepts a single channel of free surface time-series input, which is then internally converted using linear wave theory within the wavemaker control software to produce board motion.

4.1.4 Pump system

The system used consisted of a set of four pumps, each with a capacity of 0.252 m³/s (4,000 gpm) to provide both surge overtopping flow and return flow to counterbalance the mean overtopping caused by waves. Pumps were specified to provide a 0.95 m³/s discharge to generate a 0.305 m surge height. The surge height is defined as the difference between the mean water level and the elevation of the levee crest. Four 0.305-m diameter pipes were run up and over the top of the levee model, ending at the position between wave gauge array and the wavemaker.

4.2 Installation of levee-strengthening layers

The installation of the three levee-strengthening layers is described in this section.

4.2.1 Installation of HPTRM test section and maintenance

The high-performance turf reinforcement mat (HPTRM) test section is a clay-filled steel tray covered in grass that was grown into an HPTRM. The HPTRM is a vegetated slope reinforcement system that consists of a rolled-out geosynthetic composite material integrated with natural grass. The metal tray is designed to cover the crest and landward-side slope, as shown in Figure 4.7. The geosynthetic material is Profile Product Enkmat R30, and the grass is a warm-season southern Bermuda grass.

Considering the dimension of the tray (12 m long×1.8 m wide×0.7 m high) and the estimated weight of the entire system of approximately 13,000 kg, a steel tray box was used to replace the wooden box. The metal tray was designed and constructed for the installation and growth of the HPTRM system. The HPTRM system was built into a steel tray fabricated offsite by Jackson State University, located in Jackson, Mississippi, United States. The surface of the compacted clay soil was loose about 6–25 mm depth and lightly compacted.

The HPTRM roll was spread across the tray from the toe of the slope side to the toe of the crest side. Wire "U"-shaped staples fasteners were used to anchor the HPTRM mat to the compacted soil. A laboratory-sized hydro-seeder was used during the grass-seeding process. During grass growth, mechanical agitation, fertilizer, soil amendments, and mulch were used to maximize plant establishment throughout the 6-month growing season. Daily watering and weekly mowing of the grass were conducted to maintain the growth of the grass. When the full-scale overtopping tests were conducted, the vegetation had a standing thickness of 0.3 m. The steel tray, planting, and growth process of HPTRM is shown in Figure 4.8.

The HPTRM tray was covered with plastic and kept heated and illuminated during delivery to the HWRL on a flatbed trailer to prevent the glass being frosted by the cold climate in the northern United States during transportation. Once the tray arrived at

(a)

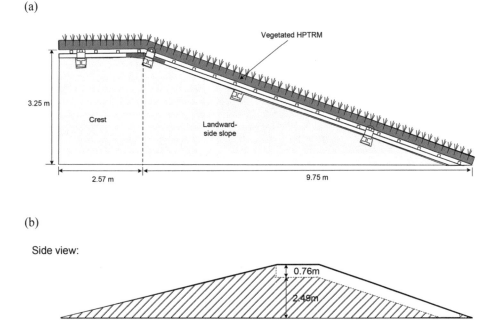

(b)

Figure 4.7 HPTRM system: (a) metal tray location in the test section, and (b) scheme of the test section.

the facility, several modifications were made to fit the tray into the test section. Next, a 110-ton crane was used to place the tray in the test section. Figure 4.9 shows the installation process of the HPTRM test section.

Prior to the start of the tests, the front of the HPTRM tray was covered with a section of galvanized steel to ease the flow transition and prevent unrealistic erosion. The sides of the tray were also covered not only to prevent erosion along the sides of the tray but also to cover the gap between the top of the tray and the edge of the test section. Finally, the toe of the tray was capped with perforated steel plates to prevent unnatural erosion of grass submerged by still water. The tray was then covered with plastic and kept warm and lit throughout the weekend and night hours to prevent the grass from becoming dormant in the relatively cold and dark environment. Maintenance had to be done to the tray because it was a living system. The system was watered every day in addition to providing additional heat and lighting throughout the hydraulic test periods. Figure 4.10 shows the maintenance process.

(a)   Seeding   (b) After 3 weeks of seeding

(c)   After 10 weeks of seeding   (d) After 5 month of seeding

Figure 4.8 HPTRM seeding and growth: (a) seeding, (b) after 3 weeks, (c) after 10 weeks, (d) after 5 months.

4.2.2 Installation of ACB test section

The scheme of the ACB test section is shown in Figure 4.11. The ACB system was a mat with 120 individual units and 30 one-and-half individual units. The one ACB block was 59 kg, and the one-and-half ACB block was 90.9 kg in weight. The individual units were staggered and interlocked for enhanced stability. Parallel strands of galvanized steel revetment cable were extending through the two cable ducts in each block allowing for longitudinal binding of the units within a mat. Each row of units was laterally offset by one-half of a block width from the adjacent row so that any given block was cabled to four other blocks (two in the row above and two in the row below).

The physical model of ACB system consisted of a single cabled-together 0.203 m thick mat placed atop a 0.102 m thick gravel filter layer over the existing compacted lean clay core inside the test section. A nonwoven geotextile with apparent opening size of 0.15 mm was placed between the gravel layer and clay, and an open plastic geogrid was placed between the gravel layer and the concrete blocks comprising the mat. Figure 4.12 shows the tapered ACB block mat.

Concrete blocks (70-T), 6 mm steel cable, geotextile fabric, and geogrid mat were provided by Contech Construction Products, Inc. Figure 4.13 shows the installation process. The geotextile cloth was cut and installed atop the compacted clay. A 3 cm

(a)   (b)

(c)   (d)

Figure 4.9 Installation of the HPTRM system: (a) HPTRM delivery and modification, (b) crane lift, (c) placement in the wave flume, and (d) final placement.

(a)   (b)

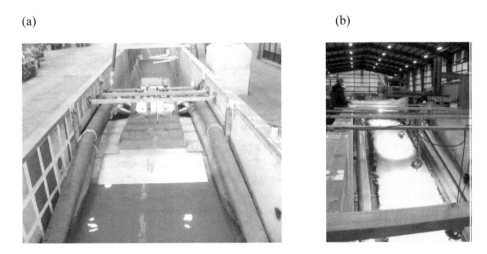

Figure 4.10 Setup of the HPTRM system in the wave flume: (a) smooth transition from the concrete section to HPTRM metal tray, and (b) overnight lighting to provide heat to prevent the grass from being dormant at low temperatures.

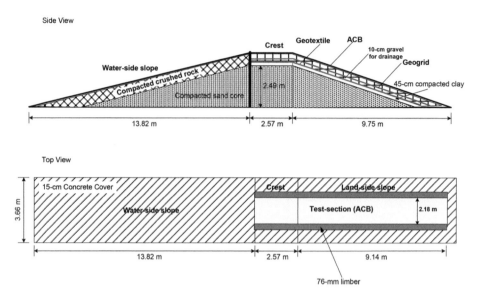

Figure 4.11 Scheme of the ACB test section.

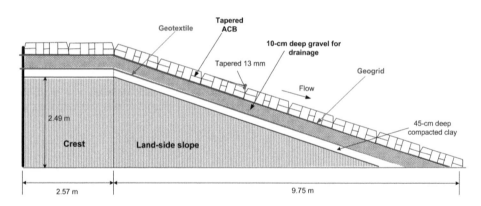

Figure 4.12 Tapered block cross-section (not to scale). The geotextile is used to separate 10-cm deep gravel from the 45-cm deep compacted clay. The geogrid is used to separate gravel from the ACB block.

open crushed rock gravel filter layer was filled to cover the geotextile and was compacted to a thickness of 0.102 m above the clay surface. The geogrid was cut to fit the area and was placed over the compacted gravel filter layer. Individual blocks were lifted into the tank using the gray cart atop the wave flume. Blocks were then transported down the landward-side slope and set on top of the geogrid, building a single row across the width of the test section. Blocks were placed from the toe up the slope to avoid putting the interlocking units in tension. A steel cable went through the blocks comprising the individual row, and then the blocks were pushed together such that adjacent blocks were touching. Because the block has a taper (1.3 cm thicker on the downslope side than upslope side), the block must be installed to ensure that there is a

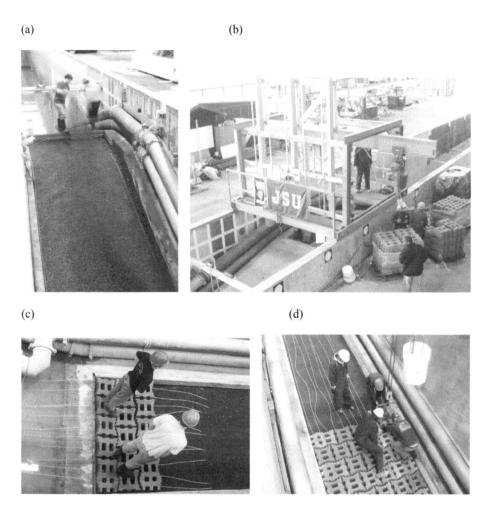

Figure 4.13 Installation of ACB system: (a) geotextile, gravel layer, and geogrid installation; (b) lifting one ACB block with a mobile crane; (c) laying down ACB
block and inserting steel cable through each block; and (d) tightening of
ACB blocks.

correct tapered block system. The final row of ACB blocks were cut to fit at the crest
and were placed along the full length of the test section. The installation of ACB was
in compliance with the manufacturer's specifications and was approved by on-site engineers from the manufacturer.

The completed physical model of the ACB mat is shown in Figure 4.14. There were
120 individual units and 30 one-and-half individual units in the ACB mat distributed
in 30 rows. After the completion of the hydraulic overtopping tests, the entire ACB
model, including concrete blocks, cables, geogrid, gravel filter layer, and geotextile
fabric, was removed. Care was taken not to disturb the underlying clay and geotechnical instrumentation. Core samples of compacted clay were obtained and geotechnical
instrumentation was recovered. The clay was excavated and hauled offsite.

(a)   (b)

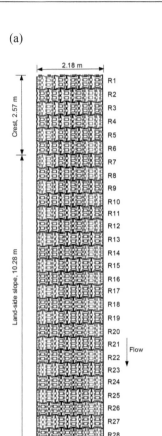

Figure 4.14 Completion for the ACB system installation: (a) scheme of ACB blocks and (b) picture of completion. There are 120 ACB blocks and 30 one-and-half ACB blocks in the system.

4.2.3 Installation of RCC test section

The scheme of the roller-compacted concrete (RCC) test section is shown in Figure 4.15. A layer of 0.3-m thick RCC was placed in a single life on top of a 0.45-m thick compacted lean clay core inside the test section. The mixing component of RCC included sand (1,050 kg/m^3), ¾ inch gravel (1,030 kg/m^3), pozzolan (270 kg/m^3), cement (60 kg/m^3), and water (50 kg/m^3). A concrete truck was used to deliver the RCC mixture from the manufacturer. A track hoe was used to place the RCC into the test section within 45 minutes of water being introduced at the mixing plant. Compaction was completed using two passes of a self-propelled, singled-drum vibratory compactor with speeds not exceeding 0.8 m/s. The resulting surface was covered by a plastic sheet and allowed to cure for 14 days. Figure 4.16 shows the construction process of the RCC test section.

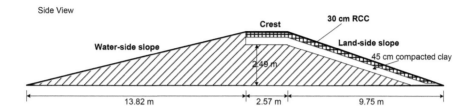

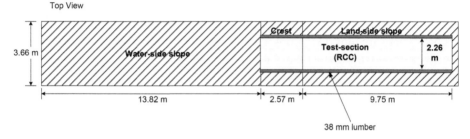

Figure 4.15 Scheme of RCC test section. (Adapted from Li et al. (2012). Reproduced with permission from Elsevier.)

(a)   (b)

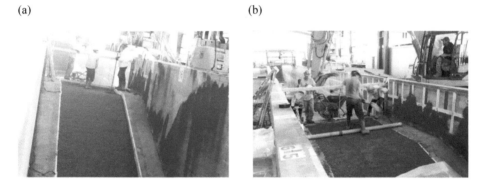

Figure 4.16 Construction of RCC test section: (a) Placement of RCC and (b) roller compaction.

After hydraulic overtopping tests in the RCC test section, the RCC layer was demolished. Care was taken not to disturb the compacted clay and instrumentation. A $0.36\,m^2$ RCC sample was collected and shipped to Oak Ridge National Laboratory for testing material properties.

4.3 Instrumentation and data collection

4.3.1 Hydraulic instrumentation

Observations made during the study included hydrodynamics and wavemaker signals. Hydrodynamic observations of free surface elevations and water particle velocities were conducted. As shown in Figure 4.17, free surface elevations were observed at up to five locations ranging from offshore of the model to partway down the water-side slope using five surface-piercing wire wave gauges (Imtech Inc.). Gauges 2–4 were placed as a three-gauge array, and gauge 1 was place not far from the array working as a backup. The distance of the three wave gauges (gauge 2–4) in the three-gauge array from the wavemaker was 28.72, 31.77 and 32.38 m, respectively. Their measurements were used to separate the incident and reflected wave and calculate the wave spectrum. Gauge 5 was placed at the end of the flume to monitor the downstream water level. The acoustic range finder (Senix TS-30S1) placed between gauge 3 and 4 was used to calibrate the wave gauges.

Water particle velocities were observed at multiple cross-shore locations atop the levee crest and down the onshore slope using eight acoustic Doppler velocimeters (ADVs). There are four down-looking ADVs and four side-looking ADVs. Both ADVs can measure water velocity up to 4 m/s. ADVs (NorTek Vectrino) were used to measure the x-component (parallel to the levee crest or slope, Figure 4.17) of flow velocity above observed locations P1–P4. The observed location P1 was located in the middle of the levee crest, P2 was located at the edge of crest and landward-side slope, and P3 and

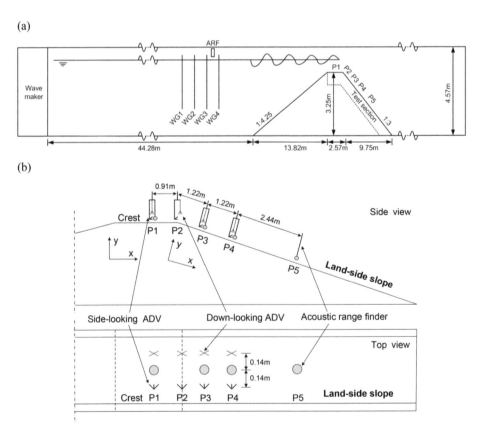

Figure 4.17 Locations of hydraulic instrumentations for the overtopping tests: (a) wave gauges and (b) ADVs and acoustic range finders. (Adapted from Li et al. (2012). Reproduced with permission from Elsevier)

P4 were located at the landward-side slope. At P5, the flow field was very disordered because of aerated flow, broken waves, and high velocity, thus no ADV was placed. ADV measures the flow velocity 5 cm away from the probe, and during measurement the probe must be submerged. When flow thickness is too small to submerge the probe, no meaningful velocity data can be collected. To get velocity data when flow thickness was small, side ADV was utilized. The measuring point of side ADV are of the same elevation on the slope. Therefore, as long as the probe can be placed, the velocity of the same elevation could be measured. At each point, side-looking ADV was fixed about 0.5 cm above the strengthening layer, measuring the flow velocity of the same height; down-looking ADV was fixed about 8 cm above the strengthening layer, measuring the flow velocity about 3 cm above the strengthening layer. An acoustic range finder was employed to measure the flow thickness at P1, P3, P4, and P5.

Water depth was observed both offshore of the levee and onshore of the levee using six acoustic range finders (Senix TS-30S1). Acoustic range finders were employed to measure the flow thickness at the observed locations P1, P3, P4, and P5. No acoustic range finder was installed at the observed location P2. Acoustic range finders measured the distance between the water surface and the probe as a function of time. The difference between the instantaneous reading and initial reading (before testing) was the instantaneous flow thickness. Figure 4.18 shows the locations of the selected wave gauges and acoustic range finders near the toe of the water-side slope, ADVs and acoustic range finders in the crest and landward-side slope section.

A wood instrument frame was built to deploy a set of eight ADVs and four acoustic range finders over the crest and landward-side slope (Figure 4.19). The frame was picked up by the gray cart atop the LWF and staged out of the way so as not to delay changes from one slope protection model to the next. The frame consisted of a wooden box beam suspended from supporting members. ADVs were attached to the sides of the box beam, with side-looking ADVs placed as close to the bed as was deemed safe. Down-looking ADVs were placed approximately 8 cm above the local bed. Acoustic range finders were mounted through the bottom of the box beam, along the center of the flume. Each set of instruments (side-looking ADV, acoustic range finder, down-looking ADV) was deployed at a common cross-shore location, with individual instrument sampling locations separated in the alongshore. All instrument locations were surveyed and recorded using the coordinate system.

Wavemaker signals observed during the tests consisted of water depth, run/stop signals used to indicate the start of the data acquisition system, observations of board motion, and free surface elevations on the wavemaker paddles. An additional voltage standard signal was recorded for quality control. Seeding material was added to provide acoustic scattering signal for the ADVs during some of the tests. This material consisted of small concentrations of neutrally buoyant hollow glass spheres (Potter Industries Sphericell™).

4.3.2 Data processing

The data acquisition system (DAQ) consists of a National Instruments that uses a real-time version of the LabVIEW programming environment. All digital and analog signals were sampled at 50 Hz, with analog signals passing through a set of built-in Butterworth anti-aliasing filters. Before each trial, the DAQ clock was set using the

(a)   (b)

(c)   (d)

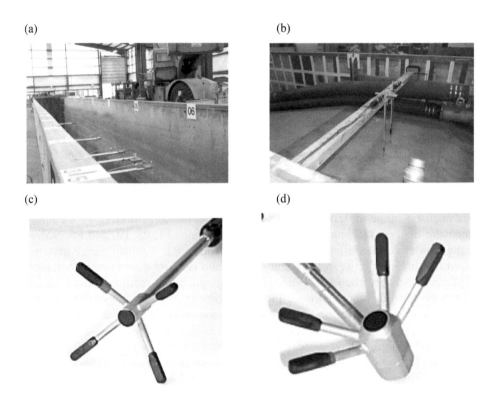

Figure 4.18 Locations of hydraulic instrumentation for the overtopping tests: (a) five surface-piercing wire gauges and one acoustic range finder along the upstream of the water-side slope, (b) four down-looking ADV, four side-looking ADV, and four acoustic range finder on the crest and landward-side slope, (c) down-looking ADV, and (d) side-looking ADV.

(a)   (b)

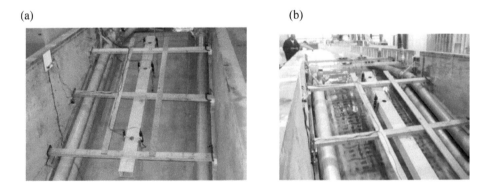

Figure 4.19 Instrumentations at (a) RCC test section and (b) ACB test section.

network time protocol against a local Stratum-2 server and three network time protocol pool servers. This resulted in a coherent array of observations with millisecond-accurate sample times. Once data collection was completed, it was copied to a secondary processor, which then copied the data onto a backed-up disk. This ensured that at least three copies were made of each data file before it was made available for postprocessing and quality control. Detailed measurement locations were kept for each trial. The instrument table locations were provided and the locations were listed in the header of data files along with information describing pertinent project, experiment, and trial information, such as the desired wave and/or surge climate and trial number.

All desired signals are subject to some unwanted signals (noise). Noise was minimized for these tests through careful controls at all stages of the DAQ, including, but not limited to, the use of shielded cabling, isolation of instrumentation power supplies from noninstrumentation outlets, elimination of ground loops, high-quality equipment from trustworthy vendors, and regular traceable calibrations. All in-situ instruments and equipment were calibrated using methods traceable to standards.

Wave gauges (Imtech Inc.) were calibrated against flume water depth during drains and fills, and dynamic calibration checks against collocated acoustic range finders were performed during postprocessing of each experimental trial. These calibration checks consisted of comparing an acoustic range finder estimate of waves to that from a nearby wave gauge during low-slope ramp-up conditions before waves became steep. Differences between the two were used to generate a rescaling parameter to correct for any changes in wave gauge calibration caused by water conductivity. Acoustic range finders (Senix TS-30S1) were calibrated prior to deployment by moving known distances away from a target and recording the distances against a calibrated and certified steel measuring tape. ADVs were calibrated at the factory (NorTek Vectrino+, 4 m/s maximum velocity, ±0.5% ±1 mm/s accuracy). The DAQ, voltage standard, and associated diagnostic equipment are calibrated yearly either by the manufacturer or by certified calibration services.

Postprocessing of all collected data was accomplished using MATLAB scripts. Sea surface elevation time series from the three-gauge array were analyzed for incident and reflected wave energy using a least-squares method of Mansard and Funke (1980), and the results were expressed in terms of energy-based incident significant wave height, H_{m0}. The other key parameters taken from frequency-domain analysis were the peak wave period, T_p, associated with the spectral peak and the mean energy wave period, $T_{m-1,0}$, determined from the resolved incident wave spectrum. The despiking of flow velocity time series from all the ADVs was conducted using mainly the method of Goring and Nikora (2002).

4.4 Testing procedures

The three physical test sections were tested under three overtopping conditions: steady flow surge-only overflow, random (irregular) wave-only overtopping, and combined wave overtopping and surge overflow. For surge-only trials, it was necessary to specify the number of pumps to be turned on. Initial tests were done with the RCC test section to determine the corresponding surge height for a particular number of pumps. For wave-only trials, the random wave properties specified were significant wave height

(H_s) and spectral peak period (T_p). For each random wave condition specified, a unidirectional TMA-type shallow water wave spectrum with standard shape factor ($\gamma = 3.3$) was generated. The generated spectrum was then used with randomly phased waveforms to generate a unidirectional, multispectral free surface time series.

4.4.1 HPTRM test section

The testing procedure for the HPTRM test section consisted of one steady surge-only overflow, followed by nine combined wave overtopping and surge overflows. The longest duration of one combined wave overtopping and surge overflow trial was up to 276 minutes. Pumps were run for a 0.296 m surge height over 1-hour duration in the surge-only overflow trial. The DAQ was started first, and then the operator would signal for the number of pumps to be turned on. This allowed estimation of the initial still water level as well as the resulting mean water level once the pumps were online and the overtopping flow had reached an equilibrium state. The resulting surge height was then estimated from the difference between the mean water level and the levee height. In this manner, the surge height corresponding to a particular number of pumps was determined for reference in subsequent experiments. The freeboard is defined as the difference between upstream water depth and crest height. The negative freeboard is for surge overflow.

For the combined waves overtopping and surge overflow trials, short runs of waves were started initially to estimate the mean overtopping rate of the waves. The procedure was similar to the surge-only overflow trials, with the DAQ starting first, followed by the wavemaker, and finally the pumps. The short trials (not shown here) were run only to determine the number of pumps it would take to dewater the onshore side of the levee during a combined wave overtopping and surge overflow condition. These trials were not run long enough for data analysis. Nine trials of the combined wave overtopping and surge overflow were run. Observations of these trials are summarized in Table 4.1, including trial number, freeboard (R_c), significant wave height (H_{m0}), peak wave period (T_p), and trial duration. Figure 4.20 shows the surge-only overflow, wave-only overtopping, and combined wave overtopping and surge overflow tests for the HPTRM test section.

Table 4.1 Combined wave overtopping and surge overflow tests for HPTRM test section

Trial number	R_c (m)	H_{m0} (m)	T_p (s)	Duration (min)
Trial 1	−0.117	0.527	6.775	90
Trial 2	−0.317	0.553	7.008	90
Trial 3	−0.307	0.649	6.826	90
Trial 4	−0.096	0.642	6.794	90
Trial 5	−0.271	0.841	6.916	90
Trial 6	−0.280	0.858	6.553	90
Trial 7	−0.285	0.868	6.863	30
Trial 8	−0.282	0.881	6.863	10
Trial 9	−0.278	0.908	7.047	276

(a)   (b)   (c)

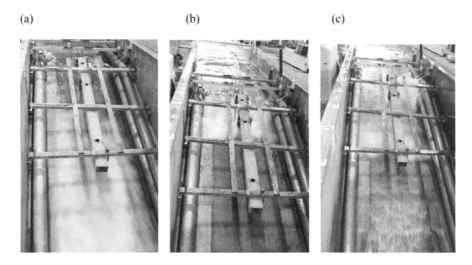

Figure 4.20 Hydraulic overtopping tests in the HPTRM test section: (a) surge-only overflow, (b) wave-only overtopping, and (c) combined wave overtopping and surge overflow.

4.4.2 ACB test section

The testing procedure for the ACB test section consisted of one steady surge-only overflow, followed by three wave-only overtopping caused by random waves, and finally four combined wave overtopping and surge overflow tests. The longest duration of one combined wave overtopping and surge overflow trial was up to 6 hours to simulate the typical storm period during hurricane conditions. Pumps were run for a 0.305-m surge height over a 4-hour duration in the surge-only overflow trial. Four trials of combined wave overtopping and surge overflow were run, and the observations of these trials are summarized in Table 4.2. Figure 4.21 shows the surge-only overflow, wave-only overtopping, and combined wave overtopping and surge overflow tests for the ACB test section.

4.4.3 RCC test section

The testing procedure for the RCC test section consisted of six different levels of steady surge-only overflow, followed by nine wave-only overtopping caused by random waves, and finally seven combined wave overtopping and surge overflow tests.

Table 4.2 Combined wave overtopping and surge overflow tests for ACB test section

Trial number	R_c (m)	H_{m0} (m)	T_p (s)	Duration (min)
Trial 1	−0.219	−0.375	4.338	90
Trial 2	−0.182	−0.282	6.473	90
Trial 3	−0.255	−0.325	4.728	90
Trial 4	−0.279	−0.373	4.728	360

(a)   (b)   (c)

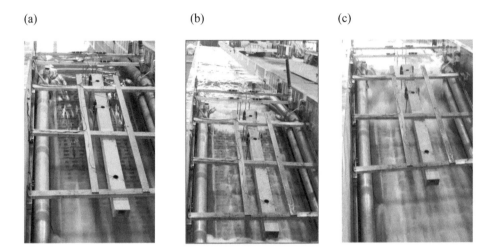

Figure 4.21 Hydraulic overtopping tests in the ACB test section: (a) surge-only overflow, (b) wave-only overtopping, and (c) combined wave overtopping and
surge overflow.

For the surge-only trials, six different overtopping rates were observed. These observations are summarized in Table 4.3. The negative freeboard is for surge overflow.
Table 4.4 summarizes the wave conditions for combined wave overtopping and surge
overflow tests in the RCC test section.

Table 4.3 Summary of surge-only overflow conditions for the RCC test section

Trial number	Freeboard R_c (m)	Duration (min)
Trial 1	−0.151	15
Trial 2	−0.148	15
Trial 3	−0.242	15
Trial 4	−0.308	15
Trial 5	−0.310	15
Trial 6	−0.388	60

Table 4.4 Combined wave and surge overtopping tests for RCC test section

Trial number	R_c (m)	H_{m0} (m)	T_p (s)	Duration (min)
Trial 1	−0.075	0.659	7.047	30
Trial 2	−0.078	0.475	3.378	15
Trial 3	−0.025	0.491	4.929	20
Trial 4	−0.017	0.568	3.432	15
Trial 5	−0.289	0.406	3.378	15
Trial 6	−0.343	0.381	4.819	20
Trial 7	−0.328	0.408	7.087	30
Trial 8	−0.321	0.496	4.682	20
Trial 9	−0.287	0.496	7.087	30
Trial 10	−0.289	0.58	5.12	20
Trial 11	−0.236	0.659	7.047	30

(a)   (b)   (c)

Figure 4.22 Hydraulic overtopping tests in the RCC test section: (a) surge-only overflow, (b) wave-only overtopping, and (c) combined wave overtopping and surge overflow.

Figure 4.22 shows the surge-only overflow, wave-only overtopping, and combined wave overtopping and surge overflow tests for the RCC test section.

4.5 Erosion check method

4.5.1 HPTRM test section

As shown in Figure 4.23, the erosion checks for HPTRM test section included 64 locations on the crest and along the landward-side slope. A hand survey of the HPTRM elevation changes was conducted for the 64 locations before any test. The elevation of the 64 stations is the initial elevation. After each test, a hand survey of the elevation changes was conducted at the 64 locations.

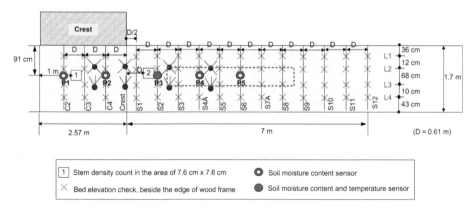

Figure 4.23 Selected locations for erosion inspection in the HPTRM test section.

Detailed grass counts within a 7.62 cm×7.62 cm wooden square at two cross-shore locations included one atop the crest between survey stations C2 and C3 and another on the landward-side slope between survey stations S1 and S2. Grass counts recorded the number of stems within each square, as well as the number of blades coming off of each stem. Figure 4.24 shows the grass counts and elevations survey.

4.5.2 ACB test section

As shown in Figure 4.25, the erosion checks for ACB test section included the following seven rows along the flow direction: R2, R5, R8, R10, R12, R14, and R17. There

(a)   (b)

Figure 4.24 Erosion inspection in the HPTRM test section: (a) hand count of grass stem and blades in an area of 7.6×7.6 cm, and (b) base elevation check.

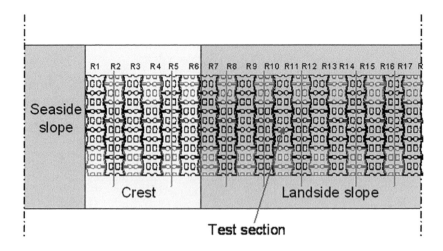

Figure 4.25 Selected locations for ACB block uplift and settlement inspection on the ACB test section.

were five blocks in each row. In the middle of each ACB block, a laser level was used to measure the initial vertical elevation. After every trial, the vertical movement (uplift or settlement) of the top of the ACB blocks was checked (Figure 4.26).

4.5.3 RCC test section

After each trial, visual inspections were performed of the RCC surface. Figure 4.27 shows the inspection location. There were eight preselected locations. At these locations, a 0.3 m×0.3 m area was checked for erosion (Figure 4.28). Visual inspections of erosion were also conducted in five large areas. Three categories of deep erosion (deeper than 12.5 mm), shallow erosion, and no erosion were used to describe the erosion.

The top layer of the RCC was found to be rough and loose, even dusty, when walked on during the setup of the instrumentation. Much of these loose materials were washed

Figure 4.26 Survey of the ACB slope protection section in the selected location (a) and laser level was used as an elevation reference (b).

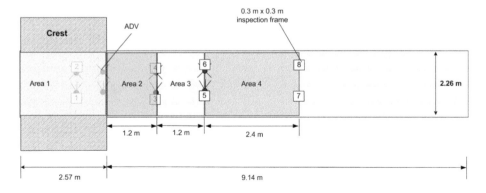

Figure 4.27 Erosion inspection on the RCC test section.

Figure 4.28 Erosion inspection box for the RCC test section (a) and severe erosion (b).

away freely during the initial tests, leaving various patches of exposed lower layers. The length, width, and depth of these eroded patches were visually inspected and photographed. Aside from these small washed-away patches, the RCC remained intact throughout all the experimental tests.

4.6 Scale, model, and measurement effects

One of the purposes of this study was to investigate the erosion characteristics of landward-side slope (including HPTRM, ACB and RCC) under combined wave and surge. Hughes (2008) pointed out that it was difficult to study the erosion characteristics of HPTRM by scale test because of the complexity of the system, so the model scale of combined wave and surge overtopping flume test was set to 1:1. Thus, the scale effects of the model did not exist in the experimental results.

The EurOtop (2016) summarized that the model and measurement effects result from the boundary conditions of a wave flume, and that measurement equipment and data analysis methods may significantly influence the comparison of results between prototype and model, or two identical models. Model and measurement effects inevitably appear in flume tests in this study, which made tests results and field data or other tests results different. The inadequate consideration of wind power, the difference between the wave spectrum used and field or other tests, the measurement error of breaking flow, and the inability to measure the flow inside the ACB block or HPTRM could lead to model and measurement effects.

Chapter 5

Testing of erosion function apparatus

After the full-scale flume tests, the high-performance turf reinforcement mat (HPTRM) system was sampled and tested using the erosion function apparatus (EFA) method (Briaud et al. 2001).

5.1 Erosion Function Apparatus (EFA)

Soil erodibility is defined as the relationship between the velocity/shear stress of the water flowing over the soil and the corresponding erosion rate experienced by the soil (Briaud et al. 2012). A critical flow velocity or critical shear stress is defined as the threshold of soil erosion. After the threshold is reached, the soil is eroded at an erosion rate that is important for researchers and engineers (Briaud and Chen 2006).

As shown in Figure 5.1, the EFA test consists of eroding a soil sample by pushing it out of a thin wall steel tube and recording the erosion rate for a given velocity of the water flowing over it (Briaud et al. 2012). During the EFA tests of both sides, 1 mm of soil is maintained out of the steel tub.

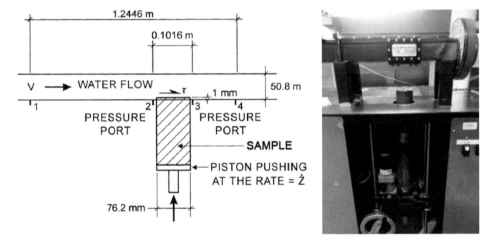

Figure 5.1 EFA adapted from Briaud et al. (2001): conceptual diagram (a); photograph of the test section (b). (After Pan et al. (2015b). Reproduced with permission from Elsevier.)

The shear stresses were estimated using the Moody chart (Moody 1944), which has been proved to be appropriate to calculate the shear stress for EFA tests (Briaud et al. 2001). In the Moody chart, the friction factor, f, can be calculated as:

$$\frac{1}{\sqrt{f}} = -2\lg\left(\frac{\varepsilon/D}{3.7} + \frac{2.51}{R_e\sqrt{f}}\right) \tag{5.1}$$

where ε/D is the pipe roughness and R_e is the Reynolds number. Using the value of friction factor, the shear stress, τ, can be calculated as:

$$\tau = \frac{1}{8}\rho f v^2 \tag{5.2}$$

where ρ is the mass density of water (1,000 kg/m^3) and v is the mean flow velocity in the pipe.

5.2 EFA tests

After the full-scale flume tests, 11 soil samples were taken from the vegetated HPTRM system: O1–O3, A1–A4, and B1–B4. The samples were 152.4 mm in length and 76.2 mm in diameter. The locations of the sampling points are shown in Figure 5.2. Before soil samples were taken, the HPTRM mat was removed because it was hard to conduct the EFA tests with the HPTRM mat on the samples. The samples had two sides, one with grass roots and another with clay only. Two of the samples (R1 and R2) were used to test the root-side, whereas the other two (C1 and C2) were used to test the clay-side for comparison.

5.3 Test results

The given velocities and corresponding shear stresses and measured erosion rates of each sample are listed in Table 5.1. The EFA test results of the samples are shown in the erodibility classification charts of velocities and shear stresses in Figure 5.3. The EFA test results of the clay-sides of the samples are plotted as hollow symbols, whereas the EFA test results of the grass-sides of the samples are plotted as solid symbols. As

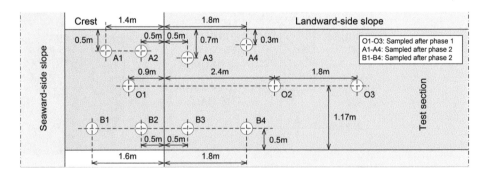

Figure 5.2 Locations of sampling points on the crest and landward-side slope. (After Pan et al. (2015b). Reproduced with permission from Elsevier.)

Table 5.1 Summary of erosion function apparatus testing results

Specimen C1			Specimen C2			Specimen R1			Specimen R2		
V	P	E	V	P	E	V	P	E	V	P	E
0.2	0.4	0.1	0.1	0.1	0.1	0.5	0.9	0.1	0.5	0.9	0.1
0.5	2.6	0.1	0.2	0.2	1.0	0.9	2.5	0.1	1.2	4.2	0.1
0.9	7.9	0.1	0.5	0.9	3.3	1.3	5.1	0.9	1.4	6.2	0.1
1.0	9.6	22.9	1.0	2.8	16.5				1.8	9.8	0.6
1.1	11.8	32.6	1.4	5.4	81.8				2.5	17.4	67.5
1.4	18.0	38.2									
1.5	23.0	68.2									

Note: *V* is applied water velocity (m/s), *P* is shear stress (Pa), and *E* is erosion rate (mm/h).

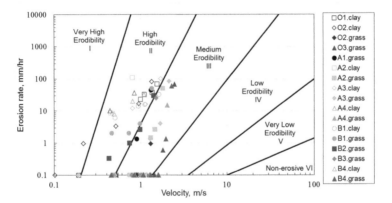

(a) Erodibility Classification Chart of Velocities

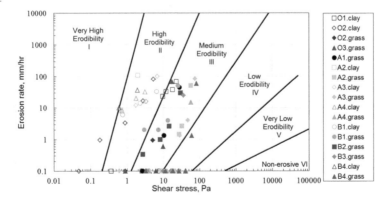

(b) Erodibility Classification Chart of Shear Stresses

Figure 5.3 EFA test results plotted in the erodibility classification charts of velocities (a) and shear stresses (b) (The erodibility classification charts are from Briaud et al. (2008)). (After Pan et al. (2015b). Reproduced with permission from Elsevier.)

shown in both the erodibility classification charts for velocities and shear stresses, the clay-sides of the samples mainly fall in Category II (high erodibility), whereas the grass-sides of the samples mainly fall in Category III (medium erodibility). The presence of grass reduces the soil erodibility from the high erodibility zone to the medium erodibility zone. Differences in the results of measurements between different samples can be explained by differences in soil properties and grass coverage quality. Figure 5.4 shows a comparison of the differences in glass cover on different samples.

The erosion rate versus shear stress curves are plotted in the linear coordinate system for all the samples (Figure 5.5). Numerous factors affect the critical shear stress and the slope of cohesive soils. Although the samples collected in this study have low liquid limits and were well-kept with the least disturbance, there are still some discrepancies in the samples. The critical shear stresses are much more consistent than the slopes. The critical shear stresses are 0.059, 2.84, and 18.50 N/m^2 for clays with no protection, HPTRM-strengthened clays with poor grass cover, and HPTRM-strengthened clays with good grass cover, respectively. Thus, even a poor grass cover can be better than the unprotected clays in increasing the critical shear stress of clay, and the critical shear stress of the HPTRM-strengthened clay with good grass cover is about six times higher than that with poor grass cover, which highlights the important contribution of full-grown grass roots to the protection of levees. The slopes vary between 5.263 and 19.231 mm/h/N/m^2 for unprotected clays, 0.178 and 0.392 mm/h/N/m^2 for HPTRM-strengthened clays with poor grass cover, and 0.031 and 0.187 mm/h/N/m^2 for HPTRM-strengthened clays with good grass cover, respectively.

The lift-up of HPTRM was observed in the EFA test for large shear stresses. However, in the experiment on the full-scale HPTRM-strengthened levee under combined storm surge overflow and wave overtopping, no lift-up or swell of HPTRM was observed with much faster overtopping flow because HPTRM was firmly interlocked with the soil as a whole.

Figure 5.4 Samples of grass cover in poor condition (AI) and good condition (A4). (After Pan et al. (2015b). Reproduced with permission from Elsevier.)

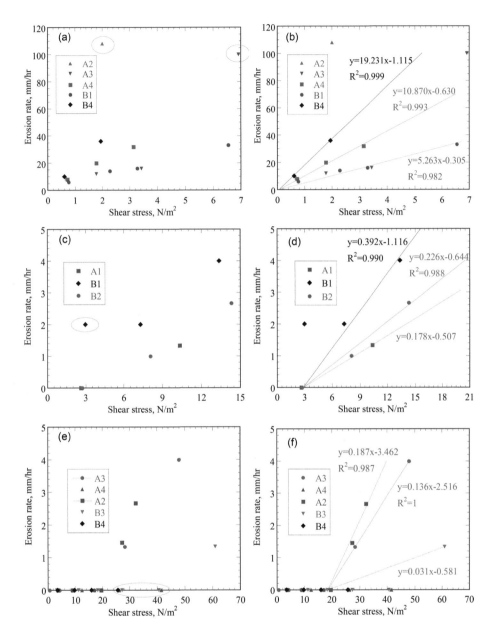

Figure 5.5 Erosion rate – shear stress plots for all samples in the linear coordinate system: (a) unprotected clays; (c) HPTRM-strengthened clay with poor grass cover; and (e) HPTRM-strengthened clay with good grass cover. Data after HPTRM's lift-up are excluded in (c) and (e) to clearly illustrate the erosion function of clay with protection. The critical shear stresses (the loadings when erosions start) are very close in each image. The average critical shear stress is used to draw best fits without circled points: (b) unprotected clays; (d) poor grass cover; and (f) good grass cover. Values of R^2 are close to 0.99 indicating very good linear relationships.

Chapter 6

Hydraulic parameters of combined wave and surge overtopping

In this chapter, the dimension analysis of the hydrodynamic parameters measured by the large-scale flume test of combined wave and surge overtopping is conducted, and the relationship between the incident wave conditions, water level, and hydraulic parameters such as the discharge across the crest and the velocity of the landward-side slope is established. A new cognitive and empirical formula is proposed to predict the hydrodynamic parameters of combined wave and surge overtopping under different water levels and wave conditions. A reference for the design, assessment, and reinforcement of levees is provided.

6.1 Distribution of incident wave

The distribution of incident waves can influence the value of the shape factor b of the Weibull distribution of the individual overtopping volumes to some degree. According to Victor et al. (2012), under positive freeboard, tests with non-Rayleigh incident waves correspond to larger values of the shape factor b. This effect also exists for the combined wave and surge overtopping, which has many similarities with overtopping. Therefore, the distribution of the incident wave is observed first.

Three examples of measured incident wave height distributions are shown in Figure 6.1. The abscissa is the ratio of the individual wave height (H) to the root-mean-square wave height (H_{rms}), the ordinates are the exceedance probability of the individual wave height, and the straight line in the figure represents the theoretical Rayleigh distribution. Figure 6.1a is an example of a good Rayleigh wave, which is measured in approximately one-fourth of all the cases. Figure 6.1b is an example of a typical non-Rayleigh wave, similar to some observation reported by Victor et al. (2012) and Nørgaard et al. (2014), showing a trend of less large waves than Rayleigh distribution. This type of distribution is measured for approximately three-fourths of all cases. Figure 6.1c is an unreasonable distribution measured in ACB test 1, showing a trend of larger waves than Rayleigh distribution. The distribution of this case is different from all other cases. The abnormal increase in the frequency of larger waves may be due to the unreasonable spectrum of the generation of the wave spectrum. The influence of this unreasonable distribution on the hydraulic parameters of combined wave and surge overtopping is shown in the following analyses.

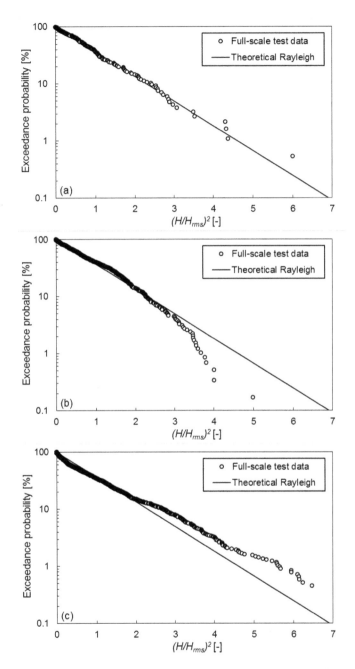

Figure 6.1 Examples of the exceedance probability distribution of incident wave height for full-scale tests. (a) Example of the cases fit well with Rayleigh distribution from test 6. (b) Example of the cases with non-Rayleigh distributed waves from test 20. (c) The case with unreasonable distribution from ACB test 1. (Adapted from Pan et al. (2015a). Reproduced with permission from Elsevier.)

6.2 Wave overtopping patterns

It was observed that the waves passing the levee crest in four different patterns during tests led to different ways of energy dissipation. Thus, it had significant effects on the levee. The four patterns are illustrated in Figure 6.2 and explained below:

1 Case 1: Plugging breakers hitting on the seaward-side slope: Waves break in front of the crest of the levee, and the impact position of the tongue is the seaward-side slope of the levee (Figure 6.2a). Before reaching the levee, when two large waves arrive one after another, the second one would usually pass in this manner. When one single mid-sized wave arrives on the crest, in most cases, it would pass in this manner too.
2 Case 2: Plunging breakers hitting on the crest: Waves break in front of the crest of the levee, and the impact position of the tongue is the crest of the levee, as shown in Figure 6.2b. When one single large wave arrives on the crest, in most cases, it would pass in this manner.
3 Case 3: Waves getting through without break: Waves directly pass the crest of the levee without a breaker, as shown in Figure 6.2c. Large waves that come in the wake of small or mid-sized waves usually go through in this this way. Single small waves or small waves that come in the wake of large waves also go through in this manner.

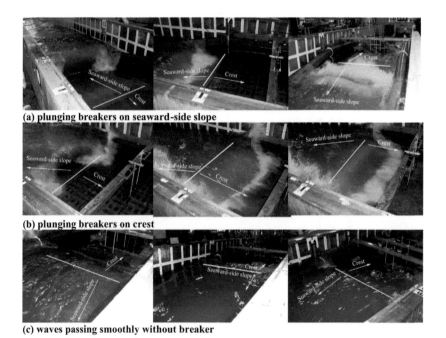

(a) plunging breakers on seaward-side slope

(b) plunging breakers on crest

(c) waves passing smoothly without breaker

Figure 6.2 Wave overtopping patterns under negative freeboard: (a) case 1, plunging breakers on the seaward-side slope; (b) case 2, plunging breakers on the crest; (c) case 3, waves passing smoothly without breaker; (d) case 4, waves passing smoothly with spilling breakers. (Adapted from Pan et al. (2015a). Reproduced with permission from Elsevier.)

4 Case 4: Surging beakers breaking on the crest (case 4): Waves directly pass the crest of the levee and with spilling breakers on the crest, as shown in Figure 6.2d. Cases (3) and (4) happen in a similar situation, but surging break occurs while the wave is running on the crest in case (4).

In cases 1 and 2, the plugging breaker and a great amount of energy dissipation occurs leading to a more broken wavefront and a turbulent mix of water flow. Thus, the waves have more erosion effects on the crest in cases 1 and 2. In cases 3 and 4, without a fierce plugging breaker, the waves have less broken wavefront, and thus large wavefront velocity occurs on the landward-side slope. Hence, the waves cause higher erosion rates on the landward-side slope in cases 3 and 4. According to the characteristic of breaking, cases 1 and 2 can be summarized as breaking passing, and cases 3 and 4 can be summarized as smooth passing. In the combined wave and surge overtopping tests 14–20 and 22, the 16-minute artificial statistics of each case was performed. The probability of each overtopping mode is listed in Table 6.1.

Figure 6.3 indicates that the breaking passing is more likely to occur with R_c/H_{m0} being close to 0, whereas smooth passing is more likely to occur with a large absolute value of R_c/H_{m0}. This can be explained by the physical meaning of relative freeboard R_c/H_{m0} under negative freeboard. Relative freeboard R_c/H_{m0} is the rate of freeboard R_c (negative) and the significant wave height based on energy spectrum H_{m0}. It reflects the proportional relationship between wave overtopping and surge overflow in the combination of wave overtopping and steady overflow. A small absolute value of R_c/H_{m0} indicates that the wave overtopping is more dominating compared to the surge overflow, and hence breaking passing is more likely to occur. A large absolute value of R_c/H_{m0} indicates that the surge overflow is more dominating compared to the wave overtopping, and hence smooth passing is more likely to occur.

To estimate the probabilities of occurrence of breaking passing and smooth passing, the recorded probabilities of occurrence are plotted versus dimensionless overtopping discharge Q_* in Figure 6.4. The Q_* is calculated according to the dimensionless overtopping (Q_B) used by Besley with q_w replaced by q_{ws}, and can be expressed as:

$$Q_* = \frac{q_{ws}}{T_m g H_s} \tag{6.1}$$

Table 6.1 Probability of each overtopping mode

Trial number	R_c/H_{m0} (−)	Case 1 (%)	Case 2 (%)	Case 3 (%)	Case 4 (%)	Breaking passing (%)	Smooth passing (%)
14	−0.345	25.49	40.69	22.55	11.27	66.18	33.82
15	−0.386	28.25	32.29	21.08	18.39	60.54	39.46
16	−0.222	51.67	7.78	6.67	33.89	59.45	40.56
17	−0.149	62.57	12.29	8.38	16.76	74.86	25.14
18	−0.705	10.71	11.90	45.83	31.55	22.61	77.38
19	−0.568	28.80	14.13	35.87	21.20	42.93	57.07
20	−0.403	37.93	25.86	22.99	13.22	63.79	36.21
22	−0.429	30.00	26.47	25.88	17.65	56.47	43.53

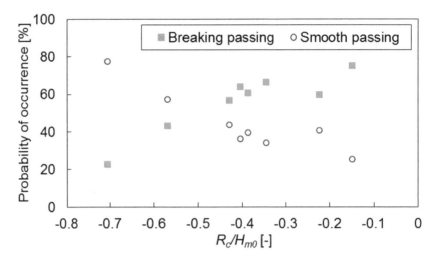

Figure 6.3 Effect of relative freeboard on wave overtopping pattern. (Adapted from Pan et al. (2015a). Reproduced with permission from Elsevier.)

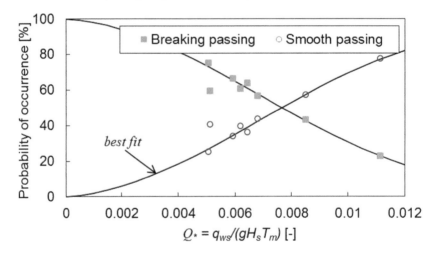

Figure 6.4 Estimates of wave overtopping patterns. (Adapted from Pan et al. (2015a). Reproduced with permission from Elsevier.)

The records give a nice trend with different Q_*, and the solid lines are the best-fit curves given by the formulas:

$$P_{sp} = 100 \tan h\left(1788 Q_*^{1.66}\right) \tag{6.2}$$

$$P_{bp} = 100 - P_{sp} \tag{6.3}$$

where P_{sp} (%) is the probability of occurrence for smooth passing, and P_{bp} (%) is the probability of occurrence for breaking passing. The applicable range of the Equations (6.2) and (6.3) is the parameter range of this experimental study.

6.3 Hydraulic parameters of surge-only overflow

Before studying the hydraulic parameters of combined wave and surge overtopping, eight surge-only overflow experiments were conducted. Hydraulic parameters such as overtopping discharge, flow thickness on the landward-side slope, and flow velocity on the landward-side slope were compared among the three levee-strengthening systems.

6.3.1 Surge-only overflow discharge

The time series of flow thickness and the time series of flow velocity at P1 in the middle of the levee crest are used to estimate the time series of surge-only overflow discharge (q_s) for each trial. Based on the measured upstream head (h_1) and steady discharges (q_s), the empirical friction (C_f) is calculated for the three types of levee-strengthening systems according to Equation (2.2). Equation (2.2) uses an empirical coefficient C_f to describe the relationship between the steady overflow discharge q_s and the upstream head h_1 (Kindsvater 1964). The steady surge-only overflow discharge (q_s) versus the upstream head (h_1) for the three levee-strengthening systems is shown in Figure 6.5. As shown in Figure 6.5, C_f is calculated by the upstream head (h_1) and steady overflow (q_s), and the Manning coefficient (n) is estimated by the Manning formula. Equation (2.2) has already been validated in the past. Thus, one ACB and HPTRM were tested to determine the empirical friction, C_f in this study. Then, it is used to estimate the corresponding steady overflow discharge according to Equation (2.2).

The empirical coefficient C_f (0.5445) of RCC tests is the same as the recommended value of Henderson (1966), while the C_f of ACB and HPTRM tests are calculated to be 0.444 and 0.415, respectively. The Manning coefficients (n) of the surface of RCC, ACB, and HPTRM are calculated to be 0.018, 0.025, and 0.035, respectively. It is noticeable that the calculation of C_f and n for ACB and HPTRM may not be precise due

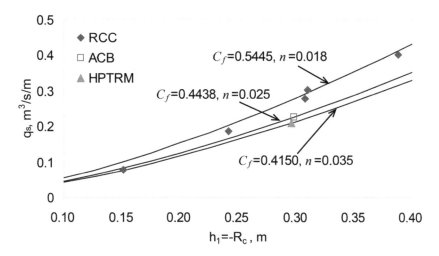

Figure 6.5 The steady surge-only overflow discharge (q_s) versus the upstream head (h_1). (Adapted from Pan et al. (2013b). Reproduced with permission from the Coastal Education and Research Foundation.)

to the lack of enough data points. However, the values of C_f and n are still meaningful for comparison. Figure 6.5 shows that the steady surge-only overflow discharge over RCC-strengthened levee is close to the existing estimation of Henderson (1966) based on laboratory tests, whereas the discharge over ACB-strengthened levee is smaller. The overflow discharge over HPTRM-strengthened levee is the smallest of the three strengthening systems.

6.3.2 Flow thickness on landward-side slope

Steady overflow on the landward-side slope of a levee is supercritical with slope-parallel velocities increasing down the slope until a balance is reached between the change of momentum of flow and the frictional resistance force of the slope surface. After the balance is reached, the characteristic values of the flow (e.g., flow thickness and flow velocity) remain constant along the landward-side slope. The average values of the time series of measured flow thickness at each measuring point were taken as the steady overflow discharge (q_s). The values of q_s at P4 and P5 were close to each other, which indicated that the balance had been reached at P5. The q_s at P5 was taken as the steady overflow discharge (q_s) on the landward-side slope. To investigate the characteristic of the steady flow thickness (d_s) on the landward-side slope a linear relationship was determined between the overflow parameter $(gd_s^3)^{1/2}$ and the steady overflow discharge (q_s), as shown in Figure 6.6. This relationship is described by

$$\frac{\sqrt{gd_s^3}}{q_s} = k_d \tag{6.4}$$

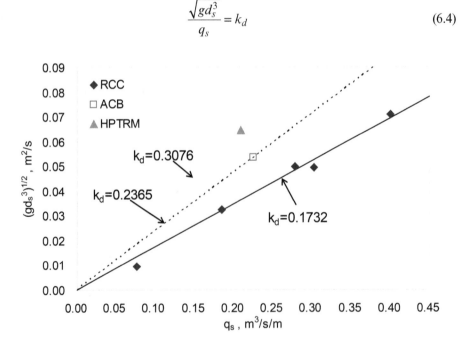

Figure 6.6 Comparison of the steady flow thicknesses (d_s) on the landward-side slope for surge-only overflow among the three strengthening systems. (Adapted from Pan et al. (2013). Reproduced with permission from the Coastal Education and Research Foundation.)

where k_d is the empirical coefficient. In this study, the best-fit k_d is 0.307 for HPTRM, 0.237 for ACB, and 0.173 for RCC. It is noticeable that the calculation of k_d for ACB and HPTRM may also not be precise due to the lack of enough data points. However, the general trends for the steady flow thicknesses (d_s) on the landward-side slope for the three strengthening systems are presented in Figure 6.6. The highest flow thickness was determined to be with the HPTRM-strengthened levee and the lowest with the RCC-strengthened levee when it was overflowed by the same discharge (q_s). This difference is also consistent with the difference in Manning coefficients (n) on the surface of the three strengthening systems.

6.3.3 Average flow velocity on landward-side slope

For surge-only overflow, the steady flow velocity (v_s) on the landward-side slope can be calculated by

$$v_s = q_s / d_s \tag{6.5}$$

Then, substituting Equations (2.2) and (6.4) into (6.5) yields

$$v_s = k_v \sqrt{gh_1} \tag{6.6}$$

with

$$k_v = \left(\frac{C_f}{k_d^2}\right)^{1/3} \tag{6.7}$$

The values of k_v are 1.64 for HPTRM, 2.00 for ACB, and 2.63 for RCC. Equation (6.6) is used to calculate the steady flow velocity on the landward-side slope of the levee. Figure 6.7 shows that the calculated values are in good agreement with the measured values.

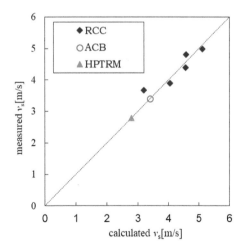

Figure 6.7 Comparison of measured steady flow velocity with calculated steady flow velocity with the three strengthening systems.

6.4 Combined wave and surge overtopping discharge

The combined wave and surge overtopping discharge is the key parameter in the design of levees and the management of coastal protection. The most representative overtopping parameter is the average overtopping discharge q_w, which is used in the design of the crest level of levees. However, according to Franco et al. (1994), distribution of individual volumes and maximum individual volume provide a better design measure than q_w, especially in the design of the protection of landward-side slope and the estimation of the effects of the wave overtopping behind the levees.

6.4.1 Combined wave and surge overtopping discharge

The time series of flow thickness and the time series of flow velocity at P1 in the middle of the levee crest were used to estimate the time series of overtopping discharge (q_{ws}) for each test. Averages of all these data were calculated for the data points collected during each test. As mentioned in Chapter 3, when flow thickness was smaller than 8 cm (i.e., the normal ADV was not submerged), the flow velocity measurement of side ADV was utilized. When flow thickness was greater than 8 cm, the flow velocity measurement of normal ADV was used.

According to Hughes and Nadal (2009), the dimensionless average overtopping discharge Q is defined as:

$$Q = \frac{q_{ws}}{\sqrt{gH_{m0}^3}} \tag{6.8}$$

Dimensionless average overtopping discharge (Q) versus relative freeboard (R_c/H_{m0}) for all the combined wave and surge overtopping tests with the three strengthening systems are shown in Figure 6.8. Several features can be seen in Figure 6.8, including:

1 The average dimensionless overtopping discharges of levee strengthened by HPTRM, ACB, and RCC are in the sequence of HPTRM < ACB < RCC for the same relative freeboard. For ACB and HPTRM tests, because of the larger Manning coefficients of the levee surface and the seepage through the interior of ACB/HPTRM, the measured overtopping discharge is less than that of RCC tests and all other estimations based on the small-scale tests with relatively smooth levee surface and no seepage through the strengthening layer.

2 The measured overtopping discharges (Q) show a strong trend with increasing relative freeboard (R_c/H_{m0}) for both RCC and HPTRM with relative freeboard $R_c/H_{m0} \leq -0.3$, but show a weak trend with relative freeboard $-0.3 < R_c/H_{m0} < 0$. This interesting phenomenon reflects the dominant problem of surge overflow and wave overtopping in the case of combined wave and surge overtopping, and distinguishing them can improve the calculation accuracy of many hydraulic parameters of combined wave and surge overtopping. This phenomenon is discussed in the following chapters in detail.

3 For the case of $R_c/H_{m0} \leq -0.3$, the dimensionless overtopping discharge (Q) on RCC can be better estimated by Hughes and Nadal (2009). The dimensionless

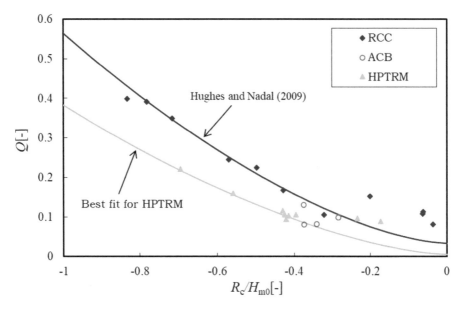

Figure 6.8 The relationship between the dimensionless overtopping discharge (Q) and relative freeboard (R_c/H_{m0}).

overtopping discharge (*Q*) on HPTRM can be calculated by the fitting formula in this study and can be expressed as follows:

$$Q = \frac{q_{ws}}{\sqrt{gH_{m0}^3}} = 0.025 + 0.37\left(\frac{-R_c}{H_{m0}}\right)^{1.75} \tag{6.9}$$

Equation (6.9) is applicable to the range of parameters in this experimental study.

4 The above analysis leaves two unsolved problems: (1) within the parameter range of $-0.3 < R_c/H_{m0} < 0$, the measured overtopping discharges (*Q*) show a weak trend with increasing relative freeboard (R_c/H_{m0}), and therefore the calculation method of dimensionless overtopping discharge (*Q*) is not given; (2) within the parameter of $R_c/H_{m0} < 0.3$, because of the lack of sufficient test groups of ACB, the estimation method of combined wave and surge overtopping discharge with ACB-strengthening layer cannot be given. These two questions will be answered through further analysis of the data in this chapter.

6.4.2 Distribution of individual overtopping volumes

Individual wave volume is an important parameter for predicting the erodibility of levee and the levee-strengthening system. Individual wave volumes are influenced by hydraulic parameters, such as overtopping discharge, significant wave height, and peak wave period.

Individual waves for all experiments were identified from the time series of water depth measured at P1. Individual waves were defined as the number of time steps from one wave when water was flowing over the levee crest. The wave period (T) of each wave was determined from the number of time steps included in each identified individual wave ($\Delta t = 0.02\,s$). Based on the starting and ending time step for each identified wave, the wave volume, was determined by integrating the calculated instantaneous discharge time series at P1 over those same time steps.

According to van der Meer and Janssen (1995) and Hughes and Nadal (2009), two-parameter Weibull distribution, as shown in Equation (2.24), was used to represent the distribution of individual wave volumes. Figure 6.9 shows an example of a good fit (a) from RCC test 10 and a mediocre fit (b) from HPTRM test 5 of the Weibull distribution to individual wave overtopping volumes. In general, for all cases, the fitting degrees between the test data and Weibull distribution are between the two cases shown in Figure 6.9.

The Weibull distribution shape factor (b) of the wave volumes distribution caused by individual waves is one of the relatively elusive parameters of the regularity. In Equation (2.41), Hughes and Nadal (2009) used two dimensionless parameters for regression analyses. The scale factor (a) of Weibull distribution reflects the size of the sample population; the shape factor (b) reflects the relative differences among individuals within the sample. Therefore, although different strengthening systems will affect the individual wave volumes, there is no longer a distinction among the three strengthening systems in this analysis. It is very difficult to find a difference between their influences on shape factor (b). On the other hand, as shown in Figure 6.8, some of the overtopping parameters show different distributions at different ranges of the relative freeboard (R_c/H_{m0}) divided by $R_c/H_{m0} = -0.3$. Thus, it was questionable whether the hydrodynamic characteristics of combined wave and overtopping are different in these two intervals. As a result of this, the groups in these two intervals were distinguished in dimensional analyses. The measured shape factor b of Weibull distribution is plotted versus dimensionless overtopping discharge $q_{ws}/(gH_{m0}T_p)$. In Figure 6.10, different symbols are used for different ranges of R_c/H_{m0}. Trends of measured b with increasing dimensionless discharge could be determined at the range of $R_c/H_{m0} < -0.3$ and $-0.3 \leq R_c/H_{m0} < 0$ separately.

The solid lines in Figure 6.10 are the best-fit curves given by the formulas:

$$b = 73.55 \left(\frac{q_{ws}}{gH_{m0}T_p} \right)^{0.76} \quad \text{for } R_c \,/\, H_{m0} \leq -0.3 \tag{6.10}$$

$$b = 54.58 \left(\frac{q_{ws}}{gH_{m0}T_p} \right)^{0.63} \quad \text{for } -0.3 < R_c \,/\, H_{m0} < 0 \tag{6.11}$$

The R-square of Equation (6.10) is 0.9749 and the root-mean-square error (RMSE) is 0.1438. On the other hand, the R-square of Equation (6.11) is 0.8526, the RMSE is 0.2065. Equations (6.10) and (6.11) apply to the range of parameters in this experimental study. It can be noted that in Figure 6.10, there is a data point (0.0063, 2.43) in the range of $R_c/H_{m0} \leq -0.3$, which deviates greatly from the fitting curve. The abnormal data may be due to the unreasonable spectrum of the generation of the wave spectrum.

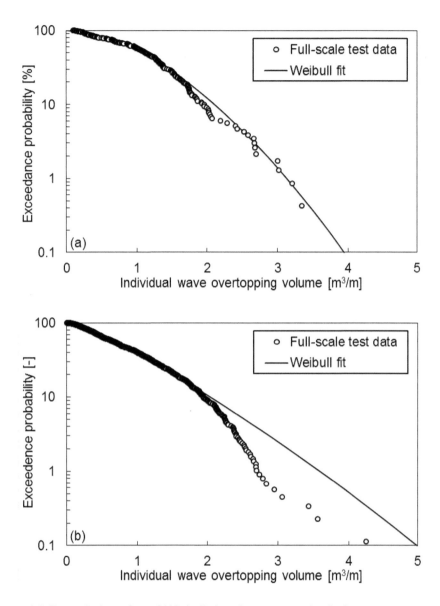

Figure 6.9 Example best fits of Weibull distribution to individual wave overtopping volume. (a) Example of a good fit from RCC test 10. (b) Example of a mediocre fit from HPTRM test 5. (Adapted from Pan et al. (2015a). Reproduced with permission from Elsevier.)

This data comes from the combined wave and overtopping tests 12, which is the only one with a typical non-Rayleigh distribution of incident wave height (smaller waves accord with Rayleigh distribution, while the frequency of larger waves is increased), as shown in Figure 6.1c.

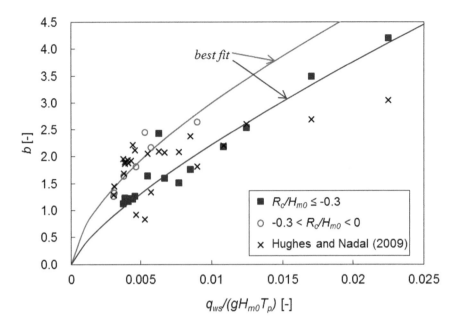

Comparing the predicted results of Equations (6.10) and (6.11) with those of Hughes and Nadal (2009), the same trend could be found in the predicted results. However, the distinction between different ranges of R_c/H_{m0} in Equations (6.10) and (6.11) gives better estimates of b, especially for the low discharge cases, in which the two distributions trends are distinguished clearly.

The scale factor a of Weibull distribution has a linear relationship with the product of the overtopping discharge and a characteristic wave period of the incident wave. In this study, the peak period T_p, the mean period T_m, and the mean (energy) wave period $T_{m-1,0}$ were examined as the characteristic wave periods. After several attempts, the mean (energy) wave period $T_{m-1,0}$ was proved to be slightly better than the peak period T_p and the mean period T_m to fit well with a. In Figure 6.11, the fitted scale factor a is plotted versus overtopping parameter $q_{ws}T_{m-1,0}$. In the range of $R_c/H_{m0}<-0.3$ and $-0.3 \leq R_c/H_{m0}<0$, the same trend for a could be found with an increase in $q_{ws}T_{m-1,0}$. The solid lines in Figure 6.11 are the best-fit curves given by the formulas:

$$a = 1.017 q_{ws} T_{m-1,0} \tag{6.12}$$

The R-square of Equation (6.12) is 0.8061 and the RMSE is 0.1654. Equation (6.12) is applicable in the range of parameters in this experimental study. The crosses in Figure 6.11 are the estimations with the equation of Hughes and Nadal (2009) given in Equation (2.4). It can be seen that Equation (2.4) fits relatively well with the data of the full-scale test, with slight underestimations.

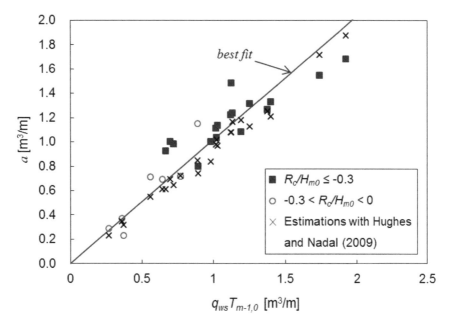

Figure 6.11 Best-fit equation for Weibull factor a. (Adapted from Pan et al. (2015a). Reproduced with permission from Elsevier.)

The mean and maximum value of the Weibull distribution can be calculated in terms of the values of the scale factor a and shape factor b as (e.g., Victor et al. 2012; Nørgaard et al. 2014):

$$b = 73.55\left(\frac{q_{ws}}{gH_{m0}T_p}\right)^{0.76} \qquad R_c / H_{m0} \leq -0.3 \qquad (6.13)$$

$$b = 54.58\left(\frac{q_{ws}}{gH_{m0}T_p}\right)^{0.63} \qquad -0.3 < R_c / H_{m0} < 0 \qquad (6.14)$$

where V_{mean} is the mean individual overtopping volumes, $V_{\max}$ is the maximum individual overtopping volumes, Γ is the gamma function, and N is the overtopping wave number.

Comparisons between estimated and measured values for mean and maximum individual overtopping volumes are shown in Figure 6.12a and b, respectively. A better prediction can be found for mean individual overtopping volumes, except for the overestimation of the two data points due to overestimated values of a (Figure 6.12). The estimation of maximum individual overtopping volumes is mediocre but reasonable. The data trend is good but the data points are scattered because of limited wavenumbers and the randomness of the maximum value in a limited number of cases tried.

It should be mentioned that in some other studies only large individual wave volumes were used to get the best fit. For instance, only the individual volumes that are larger than the average wave volume V_{mean} were used in Victor et al. (2012), whereas

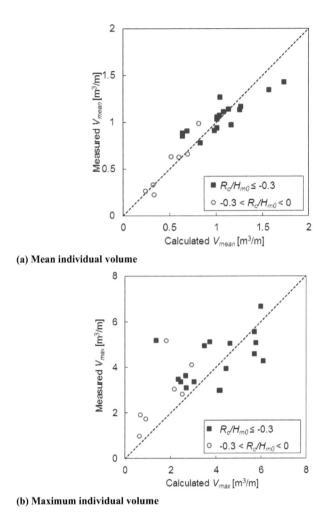

(a) Mean individual volume

(b) Maximum individual volume

Figure 6.12 Comparison of measured characteristic overtopping volumes and calculated characteristic overtopping volumes. (Adapted from Pan et al. (2015a). Reproduced with permission from Elsevier.)

only the upper 10% of the wave volumes were used in Hughes et al. (2012). The best fit with all the wave volumes (as in this study) would perform better in representing the main part of the distribution of individual volumes.

6.4.3 Distribution of instantaneous overtopping discharge

In addition to the individual wave overtopping volumes, the instantaneous overtopping discharge is also an important factor in the design and protection of levees because the peak discharge can be several times higher than the mean discharge and in many cases it leads to levee failure.

Instantaneous overtopping discharges for all experiments were identified from the time series of water depth measured at P1. According to Hughes and Nadal (2009), two-parameter Weibull distribution, as shown by Equation (2.24), was used to represent the distribution of instantaneous overtopping discharge and obtain the scale factor (a) and shape factor (b). Figure 6.13 shows an example of a good fit (a) from test 5 and a mediocre fit (b) from test 10 of the Weibull distribution to instantaneous overtopping discharges. In general, for all cases, the fitting degrees between the test data and Weibull distribution are between the two cases shown in Figure 6.13.

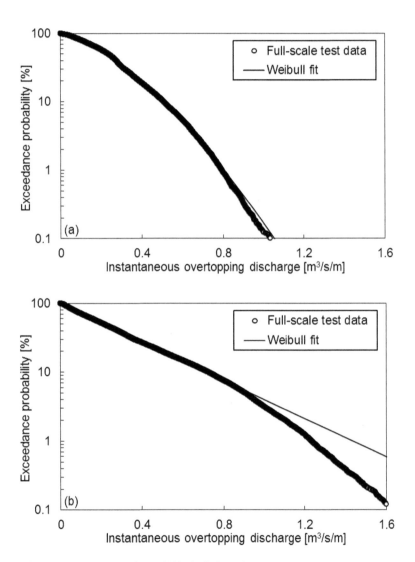

Figure 6.13 Example of best fits of Weibull distribution to instantaneous overtopping discharge. (a) Example of a good fit from test 5. (b) Example of a mediocre fit from test 10. (Adapted from Pan et al. (2015a). Reproduced with permission from Elsevier.)

After many attempts, dimensionless parameters ($q_s/(gH_{m0}T_p)$) are selected to establish the empirical relationship of shape factor (b) of Weibull distribution. The q_s is steady overflow discharge under the same freeboard calculated by Equation (2.2). It should be noted that the parameter $q_s/(gH_{m0}T_p)$ used here is different from the parameter $q_{ws}/(gH_{m0}T_p)$ used in Figure 6.10. The measured shape factor b of Weibull distribution is plotted in Figure 6.14 with dimensionless parameter ($q_s/(gH_{m0}T_p)$) as the X coordinate, and the data points in the range of $-0.3 < R_c/H_{m0} < 0$ and $R_c/H_{m0} < 0.3$ being represented by different symbols. In addition, the data points are also compared to the equation (Equation 2.43) and measurements of Hughes and Nadal (2009) derived from 26:1 laboratory tests.

It can be seen from the data points of shape factor (b) of Weibull distribution obtained from Hughes and Nadal (2009) and the full-scale tests that when parameter $q_s/(gH_{m0}T_p)$ is equal to 0, the values of fitted b are around 0.75. It is wave overtopping under zero freeboard condition and the b ($q_s/(gH_{m0}T_p)=0$) is the shape factor of the Weibull distribution of instantaneous overtopping discharge under zero freeboard condition. Therefore, in the case of overtopping, the shape factor of Weibull distribution of instantaneous overtopping discharge is close to the shape factor of Weibull distribution of overtopping caused by a single wave (see Section 2.2.3), which was around 0.75. On the other hand, the distribution of the Weibull factors b for instantaneous overtopping discharge from the full-scale tests has an obvious deviation from the estimation of Hughes and Nadal (2009), as seen from Figure 6.14. The values of b from full-scale tests are smaller than the estimation of Hughes and Nadal (2009) with increasing $q_s/(gH_{m0}T_p)$, which implies that the full-scale tests have larger values of instantaneous overtopping discharge. One possible explanation of the deviation between the results of the full-scale test and Hughes and Nadal (2009) could be the model and

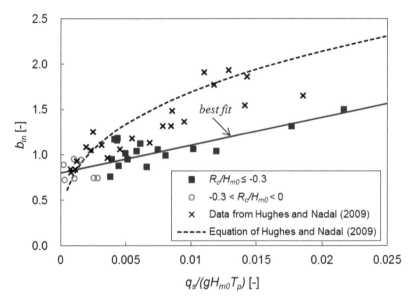

Figure 6.14 **Best-fit equation for Weibull factor b_{in}. (Adapted from Pan et al. (2015a). Reproduced with permission from Elsevier.)**

measurement techniques (see Section 3.3). The best fit of the data is given by the solid line in Figure 6.14 and Equation (6.15).

$$b = 30.64 \frac{q_s}{gH_{m0}T_p} + 0.8 \tag{6.15}$$

Equation (6.15) does not distinguish between the range of $-0.3 < R_c/H_{m0} < 0$ and $R_c/H_{m0} < 0.3$. The best-fit equation has an R-square of 0.5497 and an RMSE of 0.1242. Equation (6.14) is applicable to the range of parameters in this experimental study.

The instantaneous discharge is slightly different from that caused by individual wave volume. Its mean value is the average overtopping discharge (q_{ws}), which can be predicted by empirical formulas, such as Equations (2.39) and (6.9). Therefore, it is not necessary to fit the empirical formula for the scale factor (a) of Weibull distribution alone. According to the characteristics of Weibull distribution, having the value of Weibull factor b, the Weibull factor a can be calculated in terms of the values of the average overtopping discharge. Figure 6.15 plots the fitted a versus estimated a from Equation (6.15) for all the tests, and a better agreement is shown.

6.5 Hydraulic parameters of landward-side slope

The erosion of the landward-side slope of levees is the main reason for the failure of levees during combined wave and surge overtopping. Therefore, it is necessary to study the hydraulic parameters of the landward-side slope of levees. The overflow along the landward-side slope caused by combined wave and surge overtopping is similar to the steady overflow. Steady overflow on the landward-side slope of a levee is supercritical with slope-parallel velocities increasing down the slope until a balance is reached between the change of momentum of flow and the frictional resistance force of the slope surface. After the balance is reached, the characteristic values of the flow

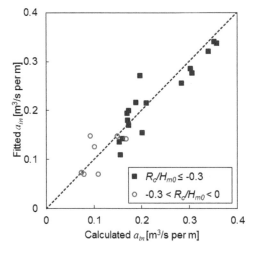

Figure 6.15 Comparison of fitted and estimated Weibull factor a_{in} from Equation (6.15). (Adapted from Pan et al. (2015a). Reproduced with permission from Elsevier.)

(e.g., flow thickness and flow velocity) remain constant along the landward-side slope. Therefore, in this research, the steady flow parameters (e.g., average flow thickness and flow velocity, RMS flow thickness and velocity, and wavefront velocity) of the flow on landward-side slope are studied. As a result, the data points at P4 or P5 (depending on different flow parameters) are selected for use in the analyses.

6.5.1 Average flow thickness and flow velocity on landward-side slope

The average flow thickness, d_m, was calculated at P5 starting with data point 1000 and continued till the end of the time series. To investigate the characteristics of the average flow thickness (d_s) on the landward-side slope, a linear relationship is determined between the average flow thickness (d_s) and the average overtopping discharge (q_{ws}), as shown in Figure 6.16. This relationship is described as:

$$\frac{q_{ws}}{\sqrt{gd_m^3}} = k_{dm} \tag{6.16}$$

where d_m is the average flow thickness, and k_{dm} is the empirical parameter related to the average flow thickness of the landward-side slope of combined wave and surge overtopping. In this study, the best-fit k_{dm} is 2.362 for HPTRM, 2.907 for ACB, and 4.811 for RCC.

If the average velocity on the landward-side slope is defined as $v_m = q_{ws}/d_m$, then

$$v_m = k_{dm}\sqrt{gd_m} \tag{6.17}$$

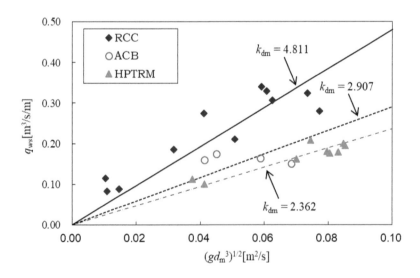

Figure 6.16 Average overtopping discharge versus average flow thickness on the landward-side slope with three levee-strengthening systems. (Adapted from Pan et al. (2013). Reproduced with permission from the Coastal Education and Research Foundation.)

According to Hughes and Nadal (2009), the Chezy equation for wide channels (hydraulic radius is approximately equal to flow thickness, d), steep slopes, and steady flow (friction slope is equal to bed slope) can be expressed as:

$$v = \sqrt{\frac{2\sin\theta}{f_F}}\sqrt{gd} \tag{6.18}$$

where θ is the slope angle and f_F is the Fanning friction factor.

To consider the influence of gradient of the slope and bottom friction, the coefficients in the two equations are equal by using Chezy equation, and the landward-side slope angle (β) is used to replace the slope angle (θ) in the Equation (6.18). Thus, the equation can be expressed as:

$$k_{dm} = \sqrt{\frac{2\sin\beta}{f_F}} \tag{6.19}$$

In this study, f_F is 0.1134 for HPTRM, 0.0748 for ACB, and 0.0273 for RCC. For the one-on-three slope in these experiments. It should be noted that the overflow caused by combined wave and surge overtopping here is not a steady flow. Thus, f_F obtained here is not the actual Fanning friction factor, and the equivalent Fanning friction factor (f_{F*}) is used instead.

Considering the influence of the gradient of the slope and bottom friction, the average flow the velocity equation becomes the following:

$$d_m = \left(\frac{q_{ws}^2 f_{F*}}{2g\sin\beta}\right)^{1/3} \tag{6.20}$$

Then, the mean flow velocity equation becomes the following:

$$v_m = \left(\frac{2q_{ws}g\sin\beta}{f_{F*}}\right)^{1/3} \tag{6.21}$$

It should be noted that Equations (6.20) and (6.21) are tentative equations and their applicability needs more experimental or field data to support.

6.5.2 Characteristic wave heights on landward-side slope

The measured time series of flow thickness at P4 were analyzed by the zero-upcrossing technique to identify the maximum and minimum flow thicknesses for all waves contained after data point 1,000. The minimum thickness was usually zero when the levee slope became temporarily "dry" during the wave trough. The following characteristic slope-perpendicular wave height parameters were determined for each measured time series: H_{rms}, $H_{1/3}$, $H_{1/10}$, and $H_{1/100}$. To test whether waves on the landward-side slope could be represented by the familiar Rayleigh distribution, estimates of the larger characteristic wave heights were made using the measured value of H_{rms} in the Rayleigh distribution formulas, as shown in Equation (6.22).

$$H_{1/3} = 1.416 H_{rms}; H_{1/10} = 1.80 H_{rms}; H_{1/100} = 2.36 H_{rms} \tag{6.22}$$

Figure 6.17 shows the comparisons between measured characteristic wave heights and those predicted by the Rayleigh distribution. The predictions for $H_{1/3}$ and $H_{1/10}$ are in good agreement with the measured characteristic wave heights. However, the prediction for $H_{1/100}$ tends to under predict slightly. A possible reason for this contradiction could be the shortness of the measured time series, resulting in only one or a few waves being averaged to estimate $H_{1/100}$. The relatively good fit of the Rayleigh distribution is helpful to characterize wave height distribution on the landward-side slope in terms of the root-mean-square wave height, H_{rms}.

Although the variation of an overflow caused by combined and surge overtopping on the landward-side slope of levees can be described by wave height distribution, for the protection of the landward-side slope of levees, more attention may be paid to the peak wave flow thickness caused by waves rather than wave height itself. The wave heights on the landward-side slope of levees show a good-fit Rayleigh distribution. Thus, the characteristic peak wave flow thickness is represented by the obtained characteristic wave heights. The measured time series of flow thickness at P4 were analyzed by the zero-upcrossing technique to identify the peak wave flow thickness. The following representative parameters of peak wave flow thickness were also determined from the measured time series: h_{rms}, $h_{1/3}$, $h_{1/10}$, and $h_{1/100}$. These values were obtained by considering only the wave peaks and not the troughs. Figure 6.18 plots ratios of four representative flow thickness parameters to the corresponding wave height parameters as a function of dimensionless factor, $R_c/(gT_p^2)$. As shown in Figure 6.18, , the ratio of representative flow thickness to representative wave height on the slope is around, when the dimensionless factor, $R_c/(gT_p^2)$, approach to 0. It is said that the "cut-off" will occur between more and more adjacent waves (i.e., the flow thickness of the valley is 0). When dimensionless factor $R_c/(gT_p^2)$ is 0, the characteristic peak flow thickness is equal to the characteristic wave height. It is wave overtopping under zero freeboard condition. In addition, different levee-strengthening systems of the landward-side slope of levees have little effects on the data distribution trend.

The solid lines shown in Figure 6.18 are best-fits to the data of surge-dominated cases expressed by the following empirical equations:

$$\frac{h_{rms}}{H_{rms}} = \exp\left[350.2\left(-\frac{R_c}{gT_p^2}\right)^{1.239}\right] \tag{6.23}$$

$$\frac{h_{1/3}}{H_{1/3}} = \exp\left[462.6\left(-\frac{R_c}{gT_p^2}\right)^{1.342}\right] \tag{6.24}$$

$$\frac{h_{1/10}}{H_{1/10}} = \exp\left[645.3\left(-\frac{R_c}{gT_p^2}\right)^{1.442}\right] \tag{6.25}$$

$$\frac{h_{1/100}}{H_{1/100}} = \exp\left[1431\left(-\frac{R_c}{gT_p^2}\right)^{1.619}\right] \tag{6.26}$$

The R-square of Equations (6.23–6.26) are 0.7845, 0.7032, 0.6376, and 0.6372, respectively, and RMSE of Equations (6.23–6.26) are 0.0241, 0.021, 0.01765, and 0.01366,

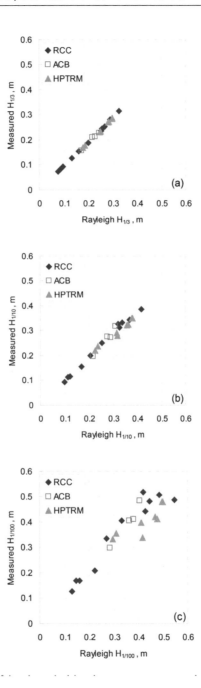

Figure 6.17 Predictions of landward-side slope $H_{1/3}$, $H_{1/10}$, and $H_{1/100}$ using measured H_{rms}. (Adapted from Pan et al. (2013). Reproduced with permission from the Coastal Education and Research Foundation.)

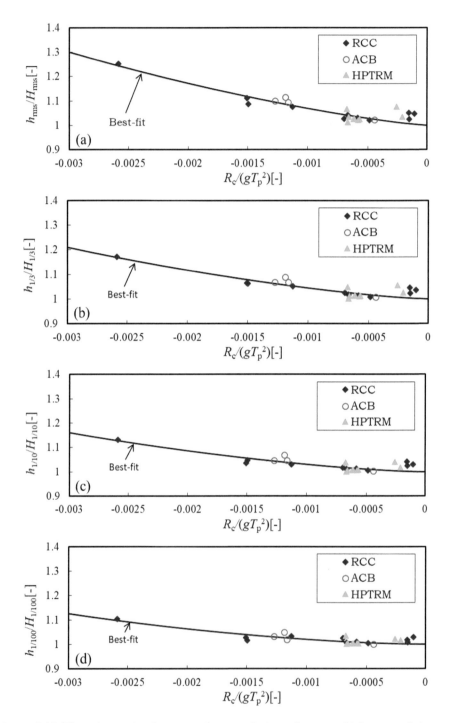

Figure 6.18 The relationship between characteristic peak water thickness and characteristic wave height on the landward-side slope of levees.

respectively. Equations (6.23–6.26) are applicable in the range of parameters in this experimental study.

6.5.3 Estimation of H_{rms} on landward-side slope

It can be seen from the previous section that the wave heights on the landward-side slope of the levees show a good-fit Rayleigh distribution during combined wave and surge overtopping, and the empirical relationship between the characteristic peak flow thickness and the corresponding characteristic wave heights can be established. Thus, if the root-mean-square wave heights (H_{rms}) on the landward-side slope are available, all kinds of characteristic wave heights and characteristic peak flow thickness on the landward-side slope can be calculated.

After many attempts, the relationship between dimensionless parameters, $H_{rms}/(-R_c)$, and relative freeboard, R_c/H_{m0}, is selected to estimate the H_{rms}. Figure 6.19 shows the relationship between dimensionless parameters, $H_{rms}/(-R_c)$, and relative freeboard, R_c/H_{m0}. The solid curve represents the best-fit of a one-parameter exponential function given as:

$$\frac{H_{rms}}{-R_c} = 0.371\left(-\frac{R_c}{H_{m0}}\right)^{0.69} \tag{6.27}$$

The R-square of Equation (6.27) is 0.967 and the RMSE of Equation (6.27) is 0.1364. Equation (6.27) is applicable in the range of parameters used in this experimental study.

6.5.4 Estimation of wave front velocity on the landward-side slope

Wavefront velocities for all experiments were identified from the time series of water depth measured at P4 and P5 using a MatLab® script. After many attempts, the

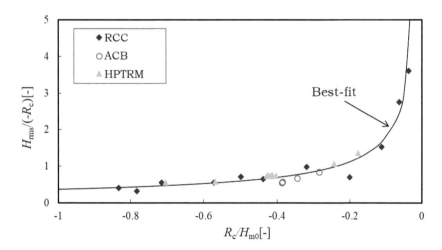

Figure 6.19 Estimation of H_{rms} on the landward-side slope as a function of R_c. (Adapted from Pan et al. (2013). Reproduced with permission from the Coastal Education and Research Foundation.)

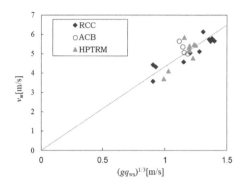

Figure 6.20 Estimated wavefront velocity determined from P4 to P5.

hydraulic parameter, $((gq_{ws})^{1/3})$ was selected to describe the wavefront velocity, v_w, on the landward-side slope. The measured front velocity, v_w, is plotted in Figure 6.20 with a dimensionless parameter, $(gq_{ws})^{1/3}$, as the X coordinate. The best empirical result for the wavefront velocity was the relationship:

$$v_w = 4.325(gq_{ws})^{1/3} \tag{6.28}$$

The R-square of Equation (6.28) is 0.6408 and the RMSE of Equation (6.28) is 0.4068. Equation (6.28) is applicable in the range of parameters used in this experimental study.

6.6 Standardized analysis of hydraulic parameters of combined and surge overtopping

Section 5.4.1 leaves two unsolved problems: one is the calculation of combined wave and surge overtopping discharge under three types of levee-strengthening systems within the range of $-0.3 < R_c/H_{m0} < 0$; and the other one is the calculation of wave and surge overtopping discharge under the surface of ACB within the range of $R_c/H_{m0} \le -0.3$.

There are two reasons for the problems mentioned above. The first is that the proportion of overtopping and overflow in the process of combined wave and surge overtopping affects its dynamic characteristics. Thus, this makes the distribution of combined wave and surge overtopping discharge with the boundary of $R_c/H_{m0} = -0.3$ to provide different characteristics. The second reason is that the distribution trend of combined wave and surge overtopping is different for different levee-strengthening systems, which needs to be studied separately. For ACB tests, it is difficult to obtain a trend because of the lack of sufficient data points.

Under the existing test data, the question becomes whether we can establish the general equivalency relationships of hydraulic parameters of combined wave and surge overtopping so that the general equivalency relationships are not affected by the characteristics of combined wave and surge overtopping itself and strengthening systems. This would not only solve the two problems that remained unsolved in Section 5.4.1 but also extends the results of this study to adapt other strengthening systems, and not just limited to the three tested strengthening systems.

Establishing such general equivalency relationships is a difficult problem. From another point of view, the surge-only overflow process is relatively simple, which can be calculated by classical hydraulic weir flow formula, and its related theoretical/empirical formula has been widely used. On the other hand, the hydraulic parameters such as overtopping discharge, average flow thickness, and average flow velocity on the landward-side slope are different for different levee-strengthening systems. Therefore, in this study, the corresponding hydraulic parameters were compared between surge-only overflow and combined wave and surge overtopping under the same freeboard (R_c). General equivalency relationships between the surge-only overflow and combined wave and surge overtopping were developed for the three studied systems or other strengthening systems. The strengthening systems were determined to have no effect on the general equivalency relationships.

6.6.1 Surge-only overflow and combined wave and surge overtopping discharge

To determine the impact of the wave action on the discharge of the combined wave and surge overtopping, the ratios of the measured combined wave and surge overtopping discharges (q_{ws}) to corresponding surge-only overflow discharges (q_s) calculated with Equation (2.2) using the same upstream heads are plotted in Figure 6.21. Figure 6.21 shows that the ratios of q_{ws}/q_s are approximately equal to 1 in the range of $R_c/H_{m0} \leq -0.3$ for all the tests of RCC, ACB, and HPTRM. For the range of $-0.3 < R_c/H_{m0} < 0$, the ratios of q_{ws}/q_s have a sharply increasing trend when R_c/H_{m0} approaches zero. Based on the trend of the ratio, it is determined that the impact of waves on the discharge of combined wave and surge overtopping is negligible in the range of $R_c/H_{m0} \leq -0.3$.

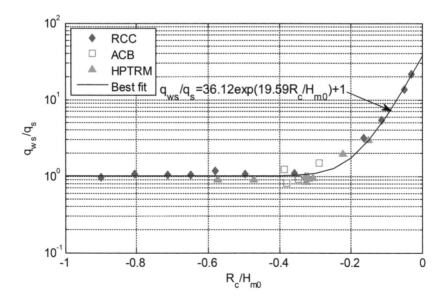

Figure 6.21 The ratio q_{ws}/q_s versus relative freeboard (R_c/H_{m0}) for the three strengthening systems. (Adapted from Pan et al. (2013). Reproduced with permission from the Coastal Education and Research Foundation.)

However, in the range of $-0.3 < R_c/H_{m0} < 0$, the discharge of combined wave and surge overtopping is mainly dominated by wave overtopping. In fact, Hughes and Nadal (2009) developed a simple relationship between combined wave and surge overtopping discharge and relative freeboard, but no further discussion was conducted. This difference takes $R_c/H_{m0} = -0.3$ as the boundary. It also explains that the combined wave and surge overtopping discharge shows different distribution trends on both sides of $R_c/H_{m0} = -0.3$, as shown in Figure 6.8. Equations (2.39) and (6.9) can only be used when the relative freeboard (R_c/H_{m0}) is large.

The best-fit curve for the ratios of q_{ws}/q_s in terms of relative freeboard R_c/H_{m0} is given as:

$$\frac{q_{ws}}{q_s} = 40.14 \exp\left(18.69 \frac{R_c}{H_{m0}}\right) + 1 \tag{6.29}$$

where q_s is calculated by Equation (2.2). Equation (2.2) fits well with the data of all the tests. In Equation (2.2), the recommended values of C_f of RCC, ACB, and HPTRM are given in Section 5.3.1. For other strengthening systems not covered in this study, the C_f of impermeable smooth surface is assumed as 0.5445, and the other strengthening systems can be tested in the laboratory on a smaller scale or estimated according to the Manning coefficient. The R-square of Equation (6.1) is 0.9949 and the RMSE is 0.349, which is applicable to the range of parameters studied in this experiment.

6.6.2 Average flow thickness on landward-side slope

Similar to the equivalency analysis of overtopping discharge, Figure 6.22 indicates that the average flow thicknesses for combined overtopping cases are higher than the corresponding steady flow thicknesses for surge-only overflow at the same discharge.

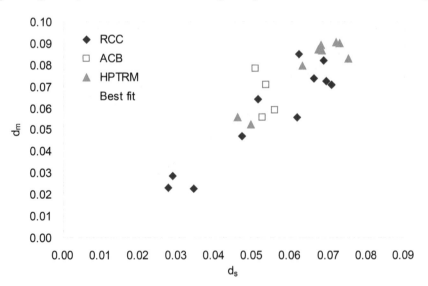

Figure 6.22 Estimation of average flow thicknesses (d_m) for combined wave and surge overtopping tests with the corresponding steady flow thickness at the same discharge.

Figure 6.22 shows the measured average flow thicknesses (d_m) from combined wave and surge overtopping tests versus calculated steady flow thickness for the same discharge using Equation (6.4). A general relationship between the average flow thicknesses (d_m) of combined wave and surge overtopping and the corresponding steady flow thickness (d_s) of surge overtopping for the three systems was observed (Figure 6.22). The straight line shown in Figure 6.22 is the best-fit curve given by the simple equation:

$$d_m = 1.174 d_s \tag{6.30}$$

where d_s is calculated by Equation (6.4) as Equation (6.4) fits well with the data of all the tests. In Equation (6.4), the recommended values of k_d of RCC, ACB, and HPTRM are given in Section 5.3.2. For other strengthening systems not covered in this study, the k_d of impermeable smooth surface is assumed to be 0.1732. Other strengthening systems can be tested in laboratory on a smaller scale or using k_d estimated according to the Manning coefficient. The R-square of Equation (6.30) is 0.8009 and the RMSE is 0.009323, which is applicable to the range of parameters studied in this experiment.

6.6.3 Average flow velocity on landward-side slope

Similar to the equivalency analysis of overtopping discharge, Figure 6.23 shows a relationship between the relative freeboard and the ratio of the measured average flow velocity of combined wave and surge overtopping (v_{ws}) to corresponding steady flow velocity (v_s) calculated with Equation (6.6) and using the same upstream heads ($h_1 = -R_c$) (k_v is the recommended value in Section 5.3.3). Similar to the distribution of ratio q_{ws}/q_s, for all tests of HPTRM, ACB, and RCC, the ratios v_{ws}/v_s are approximately equal to those in the range of $R_c/H_{m0} < -0.3$. In the range of $-0.3 < R_c/H_{m0} < 0$, the ratios of v_{ws}/v_s have a sharply increasing trend with the decreasing negative relative freeboard ($-R_c/H_{m0}$). It can be seen that the impact of waves on the average flow velocity of combined wave and surge overtopping on the land-side slope is negligible in the range of $R_c/H_{m0} < -0.3$. Within the range of $-0.3 < R_c/H_{m0} < 0$, the average flow velocity of combined wave and

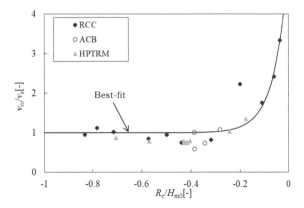

Figure 6.23 The ratio v_m/v_s versus relative freeboard (R_c/H_{m0}) for the three strengthening systems. (Adapted from Pan et al. (2013). Reproduced with permission from the Coastal Education and Research Foundation.)

surge overtopping on the land-side slope is mainly influenced by wave overtopping. The best-fit curve of the ratio v_{ws}/v_s and relative freeboard R_c/H_{m0} is expressed as:

$$\frac{v_m}{v_s} = 4.268 \exp\left(16.8\frac{R_c}{H_{m0}}\right) + 1 \tag{6.31}$$

where v_s is calculated by Equation (6.6). Equation (6.6) fits well with the data of all the tests. In Equation (6.6), the recommended values of k_v of HPTRM, ACB, and RCC are given in Section 5.3.3. For other strengthening systems not covered in this study, the k_d of impermeable smooth surface is assumed as 2.628, and the other strengthening systems can be tested in laboratory on a smaller scale or k_d estimated according to the Manning coefficient. The R-square of Equation (6.31) is 0.9081 and the RMSE is 0.1952, which is applicable to the range of parameters studied in this experiment.

6.7 Shear stress

6.7.1 Shear stress calculation

The time series of shear stress at three measurement points (P1, P3, and P4) were calculated using the momentum equation of Saint-Venant equations (Sturm 2001). For non-uniform flow on the levee slope, the external forces acting on the incremental volume are due to gravity (weight of water), the change of static pressure, and the frictional resistance of the levee slope surface material.

Considering only the one-dimensional case of a very wide channel with the major axis aligned with the levee slope, the momentum equation applicable to steep slopes can be written as (used by Nadal and Hughes 2009);

$$\frac{\partial v}{\partial t} + v\frac{\partial v}{\partial s} + g\frac{\partial h}{\partial s} + gS_f - g\sin\theta = 0 \tag{6.32}$$

or

$$S_f = \sin\theta - \frac{\partial h}{\partial s} - \frac{\partial}{\partial s}\left(\frac{v^2}{2g}\right) - \frac{\partial v}{\partial t}\cdot\frac{1}{g} \tag{6.33}$$

where S_f is the friction slope, defined as the net change in energy (head) between the two locations along a bottom boundary, θ is the angle of the levee slope to horizontal, h is the flow thickness perpendicular to the slope, v is flow velocity parallel to the slope, s is the down-slope coordinate, and t is time. In the derivation of Equation (6.32), the shear stress τ was given by Equation (6.34)

$$\tau = \gamma_w h S_f \tag{6.34}$$

where γ_w is the unit weight of water.

The calculation of shear stresses is based on Equation (6.33) and Equation (6.34). In Equation (6.33), the first term is a fixed value divided by the landward-side slope of the levee; the second term is the water surface slope; and the third and fourth relate to the spatial and temporal change of flow velocity. In this study, the time series of shear stress at P1, P3, and P4 were calculated using the time series of flow thickness and flow velocity measured at each point. Wavefront velocities on levee slope were used to estimate the changes of flow thickness and flow velocity in the down-slope coordinate for the calculation of the partial derivatives of flow thickness and the parameter $(v^2/2g)$ to the down-slope coordinate (s).

6.7.2 Shear stress analysis

The calculated time series of instantaneous shear stress were analyzed in the time domain using standard upcrossing analysis. The maximum shear stress values for each identified wave were rank-ordered, and representative values were determined for the average of the highest 1/3, highest 1/10, and highest 1/100 of the peak shear stresses. These values were denoted as $\tau_{t,1/3}$, $\tau_{t,1/10}$, and $\tau_{t,1/100}$, respectively. The comparison of characteristic shear stresses at different positions including the comparisons of $\tau_{t,1/3}$, $\tau_{t,1/10}$, $\tau_{t,1/100}$, the mean shear stress ($\tau_{t,mean}$), and the root-mean-square of the shear stress ($\tau_{t,rms}$) are plotted in Figure 6.24. As shown in this figure, the shear stresses at P3

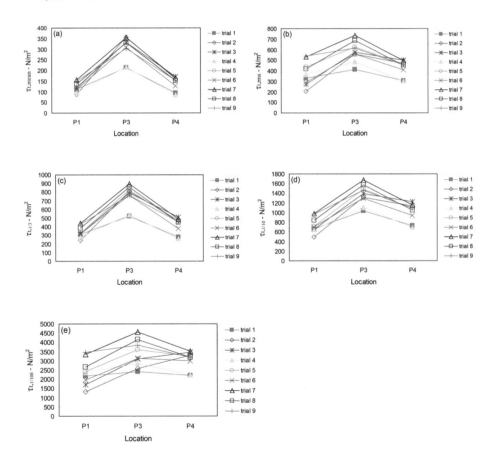

Figure 6.24 Comparison of the characteristic shear stresses at PI, P3, and P4: (a) τ_t, mean, (b) $\tau_{t,rms}$, (c) $\tau_{t,1/3}$, (d) $\tau_{t,1/10}$, and (e) $\tau_{t,1/100}$. PI is located in the middle of the levee crest. P3 and P4 are located in the landward-side slope of the levee embankment.

on the upper portion of the landward-side slope are larger than the shear stresses at P1 on the levee crest and the shear stresses at P4 on the landward-side slope.

In Table 6.2, trial 2, trial 3, and trials 5–9 were classified as surge-dominated combined wave and surge overtopping based on the relationship between average overtopping discharge and the relative freeboard. Trial 1 and trial 4 were classified as wave-dominated combined wave and surge overtopping. The trials with larger incidence wave height have larger shear stresses. It can also be seen that the mean values of the shear stresses at each position are close to that of the other trials of the same combined wave and surge overtopping type. The values of $\tau_{t,1/10}$ and $\tau_{t,1/100}$ at P4 are close to that of the other trials of this type, but at P1 and P3, these data points get dispersed.

Another finding is that although the average discharges of wave-dominated combined wave and surge overtopping cases (trial 1 and trial 4) are much smaller than those of surge-dominated combined wave and surge overtopping cases, all the characteristic shear stresses ($\tau_{t,mean}$, $\tau_{t,rms}$, $\tau_{t,1/3}$, $\tau_{t,1/10}$, and $\tau_{t,1/100}$) of wave-dominated combined wave and surge overtopping cases at P1 are larger than those of some surge-dominated combined wave and surge overtopping cases, and for $\tau_{t,1/100}$ it also happened at P3.

Based on these analyses, the influence of hydraulic condition on shear stress can be obtained. Both the overtopping discharge and the incident wave height affect the shear stresses on the levee crest and landward-side slope. The shear stress increases as the overtopping discharge and the incident wave height increase. The influence of average overtopping discharge is mainly reflected on the shear stresses on the landward-side slope. The influence of incident wave height is reflected on the shear stresses on levee crest and the top part of the landward-side slope. This is because after one wave collides with the levee and break on levee crest, it will lose most of the wave height in a short time. The larger incident wave height also makes the time series of the shear stress more uneven and causes larger $\tau_{t,1/10}$ and $\tau_{t,1/100}$ on the crest and the top part of the landward-side slope. However, at a lower position like P4, which is actually in the upper half of the levee but lower than P3, the influence of incident wave height diminishes.

Table 6.2 Hydrodynamic parameters and average wave overtopping discharge in the tests

Test number	Hydraulic parameters				Average wave/surge overtopping discharge	
	h_I	H_{m0}	T_p	$T_{m-1,0}$	q_s	q_{ws}
	(m)	(m)	(s)	(s)	$(m^3/s/m)$	$(m^3/s/m)$
Trial 1	0.117	0.527	6.775	5.648	0.052	0.100
Trial 2	0.317	0.553	7.008	5.438	0.232	0.208
Trial 3	0.307	0.649	6.826	5.618	0.221	0.200
Trial 4	0.096	0.642	6.794	5.721	0.038	0.112
Trial 5	0.271	0.841	6.916	5.597	0.183	0.181
Trial 6	0.280	0.858	6.553	5.805	0.192	0.161
Trial 7	0.285	0.868	6.863	5.631	0.198	0.195
Trial 8	0.282	0.881	6.863	5.591	0.194	0.180
Trial 9	0.278	0.908	7.047	5.635	0.189	0.176

6.7.3 Estimation of shear stresses on landward-side slope

The hydraulic conditions on the levee crest and the top part of the landward-side slope are complicated, and it is not easy to estimate the shear stresses. Steady overflow on the steep landward slope of a levee is supercritical with slope-parallel velocities increasing down the slope until a balance is reached between the momentum of the flow and the frictional resistance force of the slope surface. It is assumed that the flow down the landward-side slope caused by combined waves and surge overtopping case can also reach a similar balance after which the average flow thickness remains constant along the landward-side slope. Among all the tests, the average values of the measured time series of flow thickness at P4 and P5 are close to each other, therefore, it can be inferred that the balance has been reached at P4, and on the landward-side slope the statistical hydraulic parameters remain unchanged from P4 to the toe. Thus, the shear stress calculated at P4 can represent the shear stress condition on the landward-side slope. The shear stresses at P4 were used to estimate the shear stress on the landward-side slope. An empirical relationship is developed between the hydrodynamic parameters for each experiment and the corresponding root-mean-square of the shear stress. As indicated in Figure 6.25, the best correspondence was determined as a best-fit of the data to be a simple expression relating the root-mean-square of the shear stress to the unit weight of freshwater, γ_w, and the mean flow depth at P4, h_m, that is,

$$\tau_{t,\mathrm{rms}} = 0.0547\gamma_w h_m \tag{6.35}$$

where the mean flow depth at P4, h_m, can be replaced by the average flow thickness on landward-side slope, d_m.

Other characteristics with shear stresses including $\tau_{t,\mathrm{mean}}$, $\tau_{t,1/3}$, $\tau_{t,1/10}$, and $\tau_{t,1/100}$ can be estimated using the root-mean-square of the shear stress $\tau_{t,\mathrm{rms}}$. Figure 6.25 presents the characteristic shear stresses versus the root-mean-square of the shear stress. The solid lines in Figure 6.26 are the best-fit linear equations given as:

$$\tau_{t,1/100} = 7.04 \cdot \tau_{t,\mathrm{rms}}; \tau_{t,1/10} = 2.36 \cdot \tau_{t,\mathrm{rms}}; \tau_{t,1/3} = 0.976 \cdot \tau_{t,\mathrm{rms}}; \tau_{t,\mathrm{mean}} = 0.329 \cdot \tau_{t,\mathrm{rms}} \tag{6.36}$$

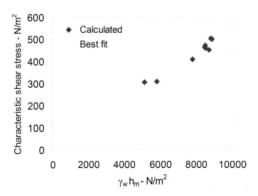

Figure 6.25 Estimation of the root-mean-square of the shear stress by the product of the unit weight of freshwater and the mean flow depth.

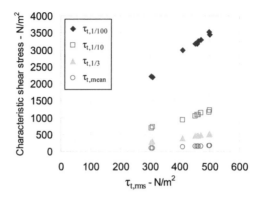

Figure 6.26 Estimation of τ_t, mean, $\tau_{t,1/3}$, $\tau_{t,1/10}$, and $\tau_{t,1/100}$ by the root-mean-square of the shear stress.

Chapter 7

Turbulent analysis

In previous chapters, the time-averaged information of overtopping flow was analyzed and explained; however, detailed flow characteristics including turbulent flow velocity and turbulent shear stress have not been discussed. Overtopping volumes of water are highly turbulent and have a substantial amount of air entrainment (Hughes et al. 2012). There is less information on the complex case of turbulent flow conditions. Turbulent flows significantly impact the design parameters for the various strengthening methods. The objective of this chapter is to understand the turbulent flow and to determine its impact on the HPTRM-strengthened levee in the combined wave and surge overtopping conditions. Despite the importance of turbulent flow in the design of a levee-strengthening system for combined wave and surge overtopping, direct measurements of overtopping turbulence are rare. Separation of turbulence from waves is difficult because turbulent velocity fluctuations typically are two to three orders of magnitude less energetic than wave motions (Shaw and Trowbridge 2001; Trowbridge and Elgar 2001; Yuan et al. 2014). This chapter presents the details of the full-scale overtopping hydraulics test, hydraulic data collection, processing, and analysis of turbulent flow. Overtopping measurements of near-bottom velocity are designed to provide direct covariance estimates of turbulent shear stress.

7.1 Measurement setup

This section describes the physical model setup. It explains the flume preparation, installation of the HPTRM systems, deployment of the instrumentation, measurement methods, and initial data analyses.

7.1.1 Model setup and instrumentation

All tests were conducted in the Large Wave Flume (LWF) at the Oregon State University's O.H. Hinsdale Wave Research Laboratory (WRL). The LWF is a 104-m-long, 3.66-m-wide, and 4.57-m-deep flume equipped with a unidirectional piston wavemaker. The physical model was set up at full scale (1:1) to eliminate errors caused by the model scale (Figure 7.1). The tested levee cross-section was built at a distance of 39.8 m from the wavemaker with a crest elevation of 3.25 m above the wave flume bottom. The sea side of the levee model section had a slope of 1V:4.25H and the land-side had a slope of 1V:3H. The large-stroke, piston-type wavemaker can generate periodic or random waves to simulate the wave spectra associated with large storms. Clean tap

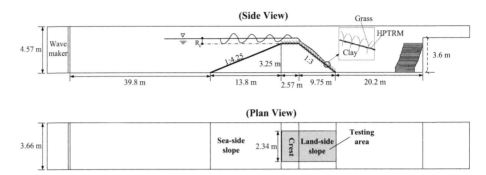

Figure 7.1 Side and profile views of the Large Wave Flume and location of levee embankment. Combined wave and surge overtopping for the embankment strengthened with HPTRM on the crest and land-side slope (*Rc* < 0) (Freeboard *Rc* is defined as the difference between the mean sea level and the elevation of the levee crest). (Adapted from Yuan et al. (2014). Reproduced with permission from the Coastal Education and Research Foundation.)

water was used in the LWF. The maximum wave height that can be produced by the LWF is 1.7 m at a peak wave period of 3.5 s in 3.5-m deep water. Because of the size and the wave-making ability of the LWF, the flume is ideally suited for levee testing under full-scale overtopping conditions.

Figure 7.2 shows the HPTRM system used during this testing. The tray was refilled with compacted clayey soils. After the clayey soil was lightly compacted, a roll of HPTRM across the tray was laid from the toe of the slope side to the toe of the crest side. Wire U-shaped staple fasteners were used to anchor the HPTRM to the compacted soil. A hydroseeder was used for uniform grass seeding on the HPTRM. The grass was a warm-season southern Bermuda grass. Fertilizer, soil amendments, and mulch were used to maximize plant establishment throughout the 5-month growing season.

To maintain the growth of the grass in the late spring and summer, water was sprayed daily and the grass was mowed weekly. Prior to delivery to the WRL using a flatbed trailer, the vegetation had a standing thickness of 0.2 m. A crane fixed the HPTRM system on to the test section, and the gap between the tray and the sidewall of the test section was sealed with a wooden enclosure. The purposes of this enclosure were (1) to ensure uniform flow as much as possible over the grass-covered portion of the land-side slope; (2) to prevent flow from flowing underneath the tray; and (3) to provide a safe walking surface during the preparation, operation, and recovery phases of the experiment.

The pump system provided both surge overtopping flow and return flow to counterbalance the mean overtopping from waves (Figure 7.3a). The pump system consisted of a set of four pumps each of which had a flow capacity of 0.252 m³/s. Pump discharge lines were run directly over the LWF wall, outside the flume, and then back into the sea-side levee slope to eliminate the disturbance to the flow going up and over the levee model.

Hydrodynamics and wavemaker signals were observed during experiments. Hydrodynamic observations were focused on free surface elevations and water particle velocities. Figure 7.3a shows the free surface elevations at six locations on the sea side

Figure 7.2 HPTRM system built for the tests.

of the levee crest. Changes in the elevations were observed using surface-piercing wire wave gauges and a single acoustic range finder. Wave gauges in the array were separated using staggered spacing (0.60, 0.92, and 2.13 m) to resolve the incident and reflected wave time series at three locations in the array.

Water particle velocities were measured at multiple cross-shore locations at the top of the levee crest and down on the onshore slope using acoustic Doppler velocimeters (ADVs) (NorTek Vectrino+, 4 m/s maximum velocity, ±1 mm/s accuracy) arranged in pairs or groups of four. Seeding material was added to provide an acoustic scattering signal for the ADVs during the flume tests. The seeding material consisted of small concentrations of neutrally buoyant hollow glass spheres (Potter Industries Sphericell). Three-dimensional (3D) water particle velocities were measured using ADVs arranged in pairs or groups of four (Figure 7.3b). Eight ADVs were placed on the crest and the land-side slope (four down-looking ADVs (ADV1–ADV4) on the middle of the crest at P1, two side-looking ADVs (ADV7 and ADV8) on the land-side slope at P3, and two side-looking ADVs (ADV5 and ADV6) near the transition between the crest and the slope at P2). The 3D water particle velocities include the streamwise (u) and cross-stream components of velocity (v), parallel to the crest or slope surface, and the vertical component of velocity (w) perpendicular to the crest or slope surface. After P4, the flow field is disordered because of aerated flow, broken waves, and high velocity, and thus no ADV was placed there. To minimize the influence of grass on ADVs, experiments were conducted before the distance between ADV and crest was gradually increased and until ADVs collected effective velocity data. Side-looking ADVs were placed within 6 cm above the HPTRM bed before the start of the experimental runs. The lowest down-looking ADV was placed 6 cm above the bed (ADV3), and subsequent ADVs were placed at higher elevations in 2 cm increments. The heights of the down-looking ADVs were wide-ranging to measure the velocity profile from the testbed to the free surface of the water.

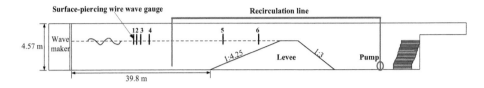

(a)

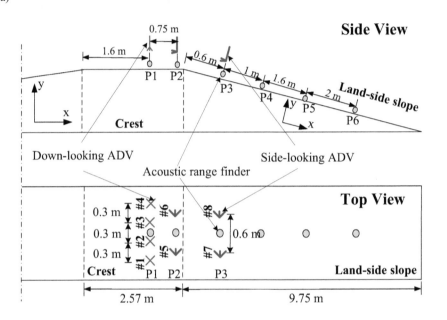

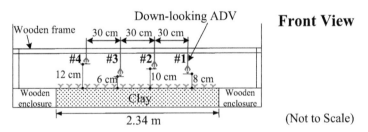

(b)

Figure 7.3 (a) Arrangement of wave gauges and pump system in the Large Wave Flume; (b) locations of ADVs and acoustic range finders on the levee model. (Adapted from Yuan et al. (2014). Reproduced with permission from the Coastal Education and Research Foundation.)

Acoustic range finders were used to measure the flow thickness at six locations (P1–P6), ranging from the levee crest and down to the land-side slope of the levee (Figure 7.3b). Acoustic range finders measured the distance between the water surface and the probe as a function of time. The difference between the instantaneous reading and the initial reading (before testing, grass lying down) was the instantaneous flow thickness.

7.1.2 Data collection and initial analyses

Eleven combined wave and surge trials with different conditions based on the capability of the piston wavemaker and the constraint of the flume wall height were designed (Table 7.1). One of the fundamental goals was to quantify the timescale over which turbulence could be considered stationary. All the hydraulic instruments were at fixed locations during all trials. Each trial was performed for 1 hour to include more than 500 individual random waves. This was statistically sufficient for stationary turbulence overtopping flow analysis. For each random wave condition specified, the system generated a unidirectional TMA-type shallow water wave spectrum with a standard shape factor ($\gamma=3.3$). The TEXEL storm, MARSEN, and ARSLOE (TMA) spectra are typical wave spectra of growing seas in limited water depths (Bouws et al. 1985). The generated spectrum was then used with randomly phased waveforms to generate a unidirectional and multispectral free surface time series. Time series of water surface elevation data and 3D flow velocity at P1–P3 were collected at a 50-Hz rate. All data were transmitted into the main control room and recorded on a computer for post-experiment processing.

Postprocessing of collected data was accomplished using MATLAB scripts. Sea surface elevation time series from the three-gauge array were analyzed for each incident and reflected wave energy using the least-squares method of Mansard and Funke (1980). The results were expressed in terms of the energy-based incident significant

Table 7.1 Hydrodynamic parameters and average wave overtopping discharge of 11 trials

Trial number	R_c (m)	H_{m0} (m)	T_p (s)	R_c/H_{m0}	q_{ws} (m^3/s-m)	SD (m^3/s-m)
1	−0.310	0.189	6.8	−1.640	0.271	0.0044
2	−0.297	0.281	6.8	−1.057	0.272	0.0036
3	−0.278	0.365	6.8	−0.762	0.253	0.0299
4	−0.248	0.456	7.1	−0.544	0.233	0.0145
5	−0.217	0.534	7.3	−0.406	0.210	0.0197
6	−0.361	0.088	6.8	−4.102	0.366	0.0036
7	−0.355	0.179	6.8	−1.983	0.362	0.0039
8	−0.348	0.276	6.8	−1.261	0.342	0.0127
9	−0.337	0.358	6.8	−0.941	0.331	0.0112
10	−0.315	0.438	6.8	−0.719	0.300	0.0113
11	−0.292	0.510	6.8	−0.573	0.256	0.0171

Note: H_{m0} = energy-based significant wave height; q_{ws} = average combined wave and surge overtopping discharge; R_c = freeboard defined as the vertical distance between the still water elevation and crest elevation; R_c/H_{m0} is the relative freeboard that reflects the proportional relationship between surge overflow and wave overtopping in the combination of surge and wave; SD = SD of the average combined wave and surge overtopping discharge; T_p = peak spectral wave period.

After Yuan et al. (2014).

wave height (H_{m0}). Another key parameter obtained from the frequency-domain analysis was the peak wave period (T_p).

Accurate ADV estimates require that the strength of the received backscattered signal must exceed the system noise. The backscattered signal strength or the signal-to-noise ratio (SNR) reported for each ADV beam is usually used for quality control in the postprocessing of the measured flow velocity. In combined wave and surge overtopping flow conditions, low signal strength may occur when the ADV sensor is not submerged during low tide or the passage of wave troughs. When the signal strength was less than 90 counts, the measured ADV data were removed and only the ADV data with signal amplitudes of 90 counts or higher were retained.

Inaccurate velocity estimates may happen when the returns from different scatters within the pairs of acoustic pulses are used to estimate the Doppler shift. Correlations between successive returns are low when excessive scatters near the sample volume reflect acoustic sidelobe energy. Thus, the along-beam correlation averaged over the sample period can be used to identify and subsequently replace potentially inaccurate data points. When the ADV data had signal correlations of less than 70%, the measured ADV data were removed and only the ADV data with signal correlations of 70% or higher were retained (Lane et al. 1998).

This study used the phase-space threshold method to remove suspected spikes in the data (Goring and Nikora 2002). The raw data collected by ADVs could not be used directly because it contained Doppler noise and ambiguity spikes. The Doppler noise is associated with the measurement process itself, which is an inherent part of all Doppler-based volume backscatter systems (Lohrmann et al. 1995). As the phase shifting between the outgoing and incoming pulses is outside the range of −180° and +180°, aliasing the Doppler signal causes a spike in the velocity record. Similar situations can also occur when the flow velocity exceeds the preset velocity range or when there is contamination from previous pulses reflected from the boundaries of complex geometries in the system (e.g., cobbles in a stream). The phase-space threshold method uses the concept of a 3D Poincare map, or phase-space plot, in which the variable and its derivatives are plotted against each other. The points are enclosed by an ellipsoid defined by a universal criterion. The points outside the ellipsoid are designated as spikes. The method iterates until the number of reasonable data points becomes constant. Interpolation between the ends of the spike was used to replace the spikes within the data.

The flow discharge was calculated using flow thickness and velocity measurements at P2 on the crest because there was relatively lower noise at this location. Figure 7.4 shows an example of the flow velocity, flow thickness, and calculated discharge time series. The raw streamwise velocity, cross-stream velocity, and vertical velocity were processed by removing weak signals (less than 90 counts signal strength), poorly correlated signals (less than 70% correlation), and despiking.

7.2 Overtopping discharge

Wave overtopping discharge rate is a critical parameter in the conceptual and preliminary design of levees. Researchers have developed several empirical formulas based on physical experiments and numerical models to predict the overtopping of levees under certain wave conditions and water levels (Pullen et al. 2007; Reeve et al. 2008; Hughes

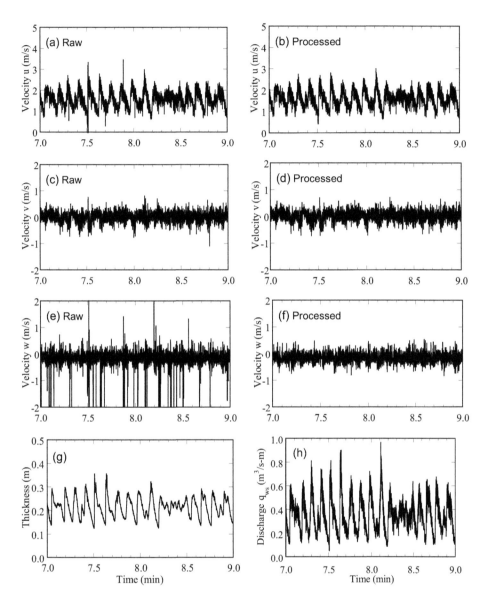

Figure 7.4 Example of flow velocity, thickness, and calculated overtopping discharge time series at P2: (a) raw streamwise velocity *u*, (b) processed *u*, (c) raw lateral velocity *v*, (d) processed *v*, (e) raw vertical velocity *w*, (f) processed *w*, (g) flow thickness, and (h) calculated discharge; ADV5 data were used, P2 is at the crest. (Adapted from Yuan et al. (2014). Reproduced with permission from the Coastal Education and Research Foundation.)

and Nadal 2009; Pan et al. 2013a, 2013b). The overtopping discharge depends on wave and structure parameters. These parameters include the seawall freeboard, crest geometry, seaward slope, significant wave height, mean or peak wave period, angle of wave incidence, water depth at the toe of the seawall, and seabed slope.

The time series of the flow thickness as well as of the flow streamwise velocity at the end of the levee crest (P2) were used to estimate the time series of the combined overtopping discharge (q_{ws}) for each trial. The time-averaged streamwise velocity and the time-averaged flow thickness from the 40th second to the end of the trials (about 1 hour) were calculated. Table 7.1 shows the calculated average overtopping discharge, its relative SD, freeboard R_c, significant wave height H_{m0}, peak wave period T_p, and relative freeboard R_c/H_{m0} for all the trials. Relative freeboard (R_c/H_{m0}) is the ratio of freeboard (R_c) to the significant wave height based on the energy spectrum (H_{m0}). It reflects the proportional relationship between surge overflow and wave overtopping in case of combined surge and wave. Relative freeboard is an important parameter in the study of wave overtopping.

Limited by the capacity of wavemaker and flume height, the highest significant wave height (H_{m0}) in all tests was fixed to 0.534 m with a corresponding peak wave period of 7.3 s and surge overflow of 0.217 m. The largest H_{m0} is near the full capability of the wave generator. The wave generator cannot generate larger waves in specified wave periods. Because of this limitation, the possible catastrophic failure of the HPTRM-strengthened levee such as the large area scour of clay with grass and slope failure of the levee could not be simulated.

Figure 7.5a shows the dimensionless combined wave and surge average overtopping discharge versus the relative freeboard for 11 trials. The dimensionless combined wave and surge average overtopping discharge are defined as $q_{ws}/\sqrt{gH_{m0}^3}$, which can be used to quantify the overtopping discharge in terms of significant wave height (van der Meer 2002; Reeve et al. 2008; Hughes and Nadal 2009). The measurements showed a good trend with increasing relative freeboard. The solid line was the best-fit empirical formula for data points as follows:

$$\frac{q_{ws}}{\sqrt{gH_{m0}^3}} = 0.0553 + 0.4765\left(\frac{-R_c}{H_{m0}}\right)^{1.58} \tag{7.1}$$

The best-fit formula had a correlation coefficient of 0.9998 and a root-mean-square (RMS) error of 0.026. According to Hughes and Nadal (2009) and Pan et al. (2013b), peak spectral wave period (T_p) had a negligible influence on determining q_{ws} at the tested range of periods. Based on the recommendations from these studies, the peak periods of the test trials were set to 7 s to obtain a larger wave force and more severe erosion on the levee. Therefore, the peak wave period was not considered when estimating q_{ws}.

According to Equation (7.1), the relative freeboard ($-R_c/H_{m0}$) dominates the relative overtopping discharge. When wave height (H_{m0}) is small and surge height (R_c) is large (i.e., $-R_c/H_{m0} > 1.0$, surge-dominated conditions), the overtopping discharge (q_{ws}) approaches surge-only overflow discharge (q_s). However, when wave height (H_{m0}) is large and surge height (R_c) is small (i.e., $-R_c/H_{m0} < 1.0$, wave-dominated conditions), the overtopping discharge (q_{ws}) is much higher than the surge-only overflow discharge (q_s). As with any empirical formula, application of Equation (7.1) is limited to the range

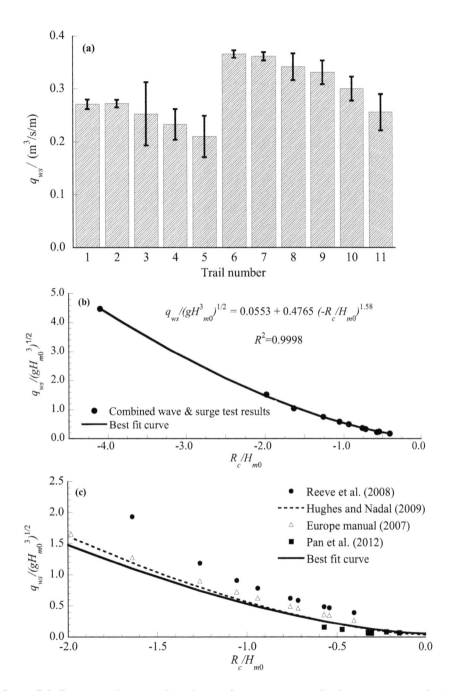

Figure 7.5 Dimensionless combined wave/surge average discharge versus relative freeboard: (a) average overtopping discharges of 11 trials (error bars represent the 95% confidence intervals ± 2 SD); (b) best-fit curve in this work; (c) comparison with the European manual (2007), Reeve et al. (2008), Hughes and Nadal (2009) and Pan et al. (2012). (Modified from Yuan et al. (2014).)

of tested parameters. In particular, seaward-side levee slopes different from 1V:4.25H could influence the wave overtopping, but seaward-side slope effects should decrease as the surge level increases.

This study used the measured wave and steady surge values to predict combined wave and surge average overtopping discharge using the empirical formulas of the European Overtopping Manual (Pullen et al. 2007; Reeve et al. 2008; Hughes and Nadal 2009). The European Overtopping Manual method and Reeve et al. (2008) use different formulas for breaking and nonbreaking waves. Hughes and Nadal (2009) developed a simple relationship between combined wave/surge discharge and relative freeboard R_c/H_{m0}.

Figure 7.5b compares the predicted overtopping discharge on a levee from this work (Equation 7.1), Pullen et al. (2007), Reeve et al. (2008), Hughes and Nadal (2009) and measured data from Pan et al. (2013b). Because of the limitations of the measurements, the data of Pan et al. (2013b) was only in the range of $R_c/H_{m0} > -0.5$. The solid line in Equation (7.1) was based on the best-fit curve of the measured data for combined wave and surge average discharge in the full-scale overtopping tests. All the prediction methods for earthen levee overestimated the measurements, whereas the formulas of Hughes and Nadal (2009) showed the closest agreement. The results from Pan et al. (2013b) were slightly lower than the estimates of the best-fit formula, which may have attributed to the flow leaking between the tested levee and the flume wall, as mentioned in Pan et al. (2013). The results in Figure 7.5b indicate that the combined wave and surge average discharge on a levee protected by HPTRM is less than the values observed for a natural earthen levee. One reason for this difference could be the higher Manning coefficient of the HPTRM.

7.3 Turbulent intensity of overtopping flow

The scouring on the embankment slope is mainly due to the joint action of fast overflow, high bed shear stress, and strong turbulent intensity. Turbulence is an important and critical factor for inducing soil erosion. The distribution of turbulence intensity in a shallow water environment is an indicator of the flow's ability to trigger and maintain sediment in suspension. This section presents turbulent characteristics including 3D velocity fluctuations and turbulent intensity. These turbulence characteristics could reflect the role of occurrence and the development of erosion on the levee crest and land-side slope.

7.3.1 Turbulence velocity fluctuations

The velocity component in the streamwise direction ignoring the effects of surface waves is $u = \bar{u} + u'$. The variance of the velocity is calculated from single-sensor velocity measurements. The fluctuation of the flow velocity is evaluated as

$$u'^2 = \text{var}(u) = \frac{1}{n} \sum_{1}^{n} (u - \bar{u})^2 \qquad (7.2)$$

where u are three components of flow velocity, $\bar{u}$ is their mean values, var(u) is variances, and n is the number of velocity data.

The velocity component in the streamwise direction is $u = \bar{u} + \tilde{u} + u'$, where the tilde denotes the wave-induced fluctuation. Equation (7.2) may not be suitable for removing wave-induced biases from estimates. Trowbridge (1998) suggested using two-sensor velocity measurements to reduce the bias produced by surface waves. This method is based on several assumptions: (1) wave-induced and turbulent velocity fluctuations are uncorrelated; (2) the statistical properties of the waves and the turbulence are stationary and ergodic; (3) and the surface waves are assumed to be weakly nonlinear and narrow-banded in both frequency and direction (Trowbridge 1998).

Two velocity sensors should be separated by a distance larger than the correlation scale of the turbulence. Then, the turbulent velocity fluctuation can be estimated by the variance of the difference of the two velocity series recorded by the two velocity sensors. The wave bias has a small magnitude which is negligible in these analyses. This study used spectra for the analyses of velocity fluctuations because it provided a suitable characterization of measurements and permitted a frequency-domain assessment for determining the stress. The calculation formula is as follows:

$$u'^2 = \text{var}(\Delta u) = 2 \int_0^{+\infty} S(f)df \tag{7.3}$$

where Δu is the time series of the simultaneous velocity minus that of the streamwise velocity, u_1 and u_2, recorded by two velocity sensors $\left(\Delta u = u^{(1)} - u^{(2)}\right)$, and superscripts (1) and (2) denote evaluation at the two ADVs in the same cross-section (e.g., ADV5 and ADV6), f is frequency, and $S(f)$ is velocity spectral density.

Figure 7.6 shows the fluctuations of the three components of flow velocity at P2 in the form of turbulent intensity. Solid cycle symbols represent the results calculated by Equation (7.2) using velocity data from ADV6, and triangular symbols represent the calculation results by Equation (7.3) using velocity data from both ADV5 and ADV6. The results calculated by Equations (7.2) and (7.3) show the difference in the prediction of the streamwise velocity fluctuation given in Figure 7.6a. It indicates that surface waves influence the estimation of turbulent velocity fluctuation. The streamwise velocity fluctuation $\left(u'^2\right)$ was always the highest among the three velocity fluctuations at P2 or P3 for all 11 trials.

Figure 7.7 shows the comparisons of energy spectrum densities of u'^2, v'^2, and w'^2 using the single-sensor method (Equation 7.2) and the two-sensor method (Equation 7.3). According to Figure 7.7a, c and e, a significant amount of energy is located at the wave peak frequency (0.15 Hz), especially for the energy spectrum densities of u'^2. However, no such obvious energy concentration near the wave peak frequency exists in the two-sensor method (Figure 7.7b, d and f), which suggests that the Trowbridge (1998) method could effectively remove the effects of surface waves on turbulent velocity calculations. Surface waves have more influence on the streamwise velocity fluctuation, u'^2, than the velocity fluctuations in the other two directions. The difference in the time series of velocity data recorded by the two separate ADVs could reduce the wave bias. Therefore, all velocity turbulent intensities $\left(u'^2, v'^2, \text{and } w'^2\right)$ in the following sections were obtained using the Trowbridge (1998) method.

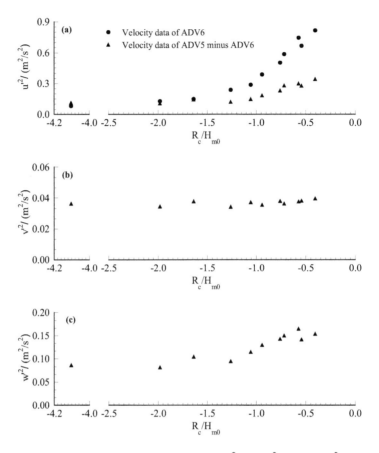

Figure 7.6 Calculated turbulent intensities of (a)u'^2, (b) v'^2, and (c)w'^2 using velocity data collected by ADV6. The u'^2 was also calculated using velocity data collected by ADV5 and ADV6. (Adapted from Yuan et al. (2014). Reproduced with permission from the Coastal Education and Research Foundation.)

7.3.2 New formula for turbulent intensity on the crest and the land-side slope

The absolute intensity of velocity fluctuations (variances) can be used to infer the bed stress through the turbulent kinetic energy as follows:

$$k = \frac{1}{2}\left(u'^2 + v'^2 + w'^2\right) \qquad (7.4)$$

Equation (7.4) estimated velocity fluctuations $\left(u'^2, v'^2, \text{and } w'^2\right)$ to reduce the bias produced by the surface waves. Empirical correlations were developed between the turbulent kinetic energy ($k/g/(-R_c)$) and relative freeboard ($-R_c/H_{m0}$) at P2 and P3. These correlations indicated that there was a critical value of $-R_c/H_{m0}=2.0$ for the turbulent kinetic energy ($k/g/(-R_c)$). Figure 7.8 shows that the measurements have a good trend.

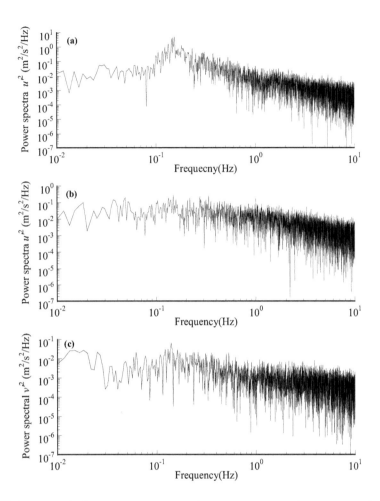

Figure 7.7 An example for energy spectrum densities using single-sensor methods for (a) u'^2; (c) v'^2; (e) w'^2 and two-sensor method for (b) u'^2; (d) v'^2; (f) w'^2 in the single-sensor method are from ADV5; and in the two-sensor method are from ADV5 and ADV6 in trial 9. (Modified from Yuan et al. (2014).)

Relative freeboard $(-R_c/H_{m0})$ increased when $-R_c/H_{m0} < 2.0$ both at P2 and P3. For $-R_c/H_{m0} > 2.0$, $k/g/(-R_c)$ tended to constant. The turbulent kinetic energy (k) has a linear relationship with the freeboard $(-R_c)$, and the wave is negligible when $-R_c/H_{m0} > 2.0$. The turbulent kinetic energy (k) at P3 was higher than that of P2 because there was more turbulent flow on the land-side slope. Equation (7.5) gives the best-fit formulas for the turbulent kinetic energy (k) at P2 with a correlation coefficient of 0.945 and an RMS error of 0.017. Equation (7.6) gives the best-fit formulas for the mean turbulent intensity at P3 with a correlation coefficient of 0.952 and an RMS error of 0.081.

$$\frac{k}{g(-R_c)}\bigg|_{\text{at P2}} = \begin{cases} 0.0563\left(-R_c/H_{m0}\right)^{-0.83}, & -R_c/H_{m0} < 2.0 \\ 0.0328 \quad, & -R_c/H_{m0} > 2.0 \end{cases} \tag{7.5}$$

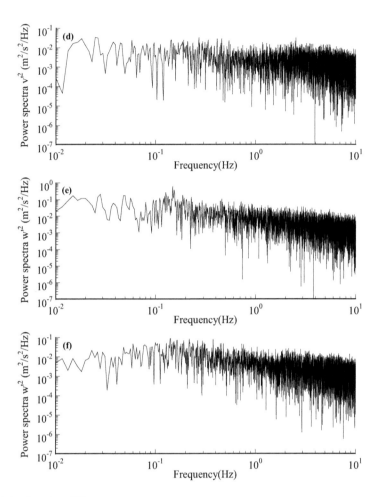

Figure 7.7a Caption TK

$$\frac{k}{g(-R_c)_{\text{at P3}}} = \begin{cases} 0.263\left(-R_c/H_{m0}\right)^{-0.89} & , \quad -R_c/H_{m0} < 2.0 \\ 0.1442 & , \quad -R_c/H_{m0} > 2.0 \end{cases} \tag{7.6}$$

Similar to Equation (7.1), the application of Equations (7.5) and (7.6) is limited to the range of tested parameters.

7.4 Turbulent shear stress

The bed shear stress is a fundamental variable in flow studies to determine the transport field, scour, deposition, and bed change (Wilcock 1996). Five methods including the log profile method, the Reynolds stress method, the turbulent kinetic energy (k) method, the turbulent kinetic energy (w') method, and Nadal and Hughes' method

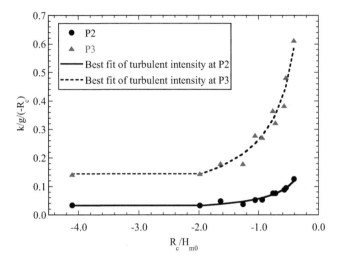

Figure 7.8 Dimensionless turbulent intensity $k/g/(-R_c)$ at P2 and P3 versus relative R_c/H_{m0}. (Adapted from Yuan et al. (2014). Reproduced with permission from the Coastal Education and Research Foundation.)

were used for turbulent shear stress estimations. The calculated turbulent shear stress, determined by the five methods, was compared to determine the most suitable method for the combined wave and surge overtopping analyses.

7.4.1 Log profile method

The log profile estimates of stress are common in coastal research (Trowbridge 1998). Four down-looking ADVs (ADV1–ADV4) were set at P1 on the crest to obtain the velocity profile for this method. ADV4 was placed 12 cm above the bed surface, which was the highest location among the four down-looking ADVs. During the 11 trials, ADV4 produced meaningless data when the water elevation was beneath the probe. When this occurred, the data from this ADV was discarded. The study used the data from ADV1, ADV2, and ADV3, which were placed with sampling volumes of 8, 10, and 6 cm, respectively, above the local bed surface to analyze the velocity profile.

In the log profile method for the grass-covered bed surface, the local bed shear stress (τ_0) is calculated from the logarithmic relation between the shear velocity and the variation of velocity with height (Stephan and Gutknecht 2002) as

$$\frac{u}{u_*} = \frac{1}{\kappa} \ln\left(\frac{z - z_{p,m}}{z_{p,m}}\right) + 8.5 \tag{7.7}$$

where u is velocity, u_* is shear velocity $\left(\sqrt{\tau_0 / \rho}\right)$, ρ is water density, κ ($= 0.41$) is von Kármán constant, z is the height above the bed, $z_{p,m}$ is the mean plant height, and the integration constant for a rough bed is 8.5 (Christensen 1985; Stephan and Gutknecht 2002).

Approximately 56 minutes (from the 4th to 60th minute) of velocity data were averaged to obtain the mean streamwise velocity at the same cross-section of P1 (Table 7.2).

The analysis is based on the assumption that velocity fluctuations with time scales of less than 1 hour that are created by superposed surface waves and turbulence (Trowbridge and Elgar 2003). In the Reynolds shear stress $\left(-\rho\overline{u'w'}\right)$, the null hypothesis is that the dynamics of turbulence in the shallow coastal ocean, at heights above bottom, are small relative to the water depth, but larger than the thickness of the oscillatory boundary later produced by surface waves. The hypothesis is consistent with the local application of Monin–Obukhov scaling (1971). In this scaling, the statistical properties of the velocity only depend on z, ρ, the vertical buoyancy flux, and the vertical flux of streamwise momentum (approximated by $\rho\overline{u'w'}$). The assumption in the log profile method also includes the constancy of stress in the turbulent boundary layer. Equation (7.7) was used to calculate the friction velocity (u_*). The deflected grass height $(z_{p,\,m})$ was measured during the experiments. When the wavemaker stopped and the levee embankment dried after each trial, the grass height was measured immediately at P1. Grass heights during the 11 trials were similar, averaging 1.25 cm. Three friction velocities were measured using three mean streamwise velocities at ADV1, ADV2, and ADV3. The average of these three shear velocities was noted as the mean shear velocity. Then, the turbulent shear stress (τ) is estimated by

$$\tau_0 = \rho u_*^2 \tag{7.8}$$

where ρ is the density of water. Table 7.2 lists the calculated mean streamwise velocity of the three ADVs and the mean turbulent shear stress at P1. The mean turbulent shear stress is in the range of 11.53 and 14.07 N/m^2.

7.4.2 Reynolds stress and turbulent kinetic energy methods

The statistical properties of the flow are assumed to be stationary over the averaging period and independent of the ADV position. The quantity $-\rho\overline{u'w'}$ is evaluated at a

Table 7.2 Mean streamwise velocity and mean turbulent shear stress measured by ADV 1, ADV 2, and ADV 3 for all 11 trials

Trial number	R_c/H_{m0}	Mean streamwise velocity u (m/s)			Mean turbulent shear stress τ (N/m^2)
		ADV#1	ADV#2	ADV#3	
1	−1.640	1.338	1.491	1.275	11.70 ± 0.19
2	−1.057	1.446	1.516	1.344	12.90 ± 0.27
3	−0.762	1.334	1.559	1.304	12.23 ± 0.17
4	−0.544	1.430	1.531	1.335	12.83 ± 0.29
5	−0.406	1.257	1.552	1.268	11.53 ± 0.47
6	−4.102	1.462	1.608	1.384	13.78 ± 0.37
7	−1.983	1.448	1.550	1.352	13.15 ± 0.32
8	−1.261	1.459	1.560	1.361	13.34 ± 0.29
9	−0.941	1.566	1.567	1.414	14.39 ± 0.37
10	−0.719	1.523	1.557	1.389	13.89 ± 0.09
11	−0.573	1.508	1.593	1.398	14.07 ± 0.24

Note: The value after ± is the SD.

After Yuan et al. (2014).

height above bottom that is small compared with the water depth, but large relative to the scale of the bottom roughness elements. Thus, it represents the stress transmitted to the HPTRM floor. The stress-transmitting turbulence is assumed to be distinguishable from other motions because of its small spatial scales, which is believed to be a small fraction of the wave depth (e.g., Svendsen and Putrevu 1994).

In the Reynolds stress method, the local bed shear stress is determined from the Reynolds stress when the turbulence measurements are available as follows:

$$\tau(z) = -\rho \overline{u'w'} \tag{7.9}$$

where $\tau(z)$ is shear stress, u' and w' are the velocity fluctuations of the streamwise and vertical components, and the overbar denotes an average (Pope 2000; Babaeyan-Koopaei et al. 2002).

Local turbulent shear stress can be determined from Reynolds stress using Equation (7.9) when turbulence measurements are available. As discussed earlier, the use of single-sensor velocity measurements may not be feasible for removing wave-induced biases from the estimates of turbulent shear stress $(-\rho \overline{u'w'})$. The turbulent shear stress is usually estimated by $\rho \cdot \mathrm{cov}(u,w)$, where cov() represents the covariance. In this analysis, the Reynolds stresses were also calculated using two-sensor velocity measurements to reduce the bias produced by surface waves. Reynolds stress was calculated by cov(u, w) for single velocity sensors with the error produced by surface waves. For two velocity sensors method, the Reynolds stress was calculated by $\mathrm{cov}(\Delta u, w^{(1)})$, where Δu was the time series of the simultaneous difference of streamwise velocity u_1 and u_2, recorded by two velocity sensors.

The use of spectra provides a suitable characterization of the measurements and permits a frequency-domain assessment of the technique for determining the stress. It is convenient to let S_{ij} be the real part of the cross-spectrum of Δu and $w^{(1)}$ to present spectra, defined as:

$$\overline{u_i'u_j'} = \mathrm{cov}\left[\Delta u_i, u_j^{(1)}\right] = 2 \int_0^{+\infty} S_{ij}(f)\,df \tag{7.10}$$

Figure 7.9 shows an example of the comparison of cross-spectra spectrum density using the single-sensor method and the two-sensor method. The co-spectrum of u and w, both collected by ADV5, indicated an energetic and narrow peak which clearly coincided with the wave peak. The co-spectrum of w from ADV5 and $\Delta u(u_1 - u_2)$ based on the difference between ADV5 and ADV6 velocities was an order of magnitude less energetic at its peak and much broader. This observation suggests that the Trowbridge (1998) method reduces the bias produced by surface waves. However, because of the limitation of the method itself, the method only minimized the influence of surface waves and did not eliminate it completely, which yielded some energy concentration (Figure 7.9b). With the current wave research progress, it is impossible to entirely eliminate the influence of surface waves.

Linear relationships between turbulent energy and shear stress have been developed in a quasi-equilibrium turbulent energy model (e.g., Galperin et al. 1988). Soulsby and Dyer (1981) analyzed turbulence data of tidal currents and showed that the average ratio of shear stress to turbulent kinetic energy is constant. Soulsby (1983) extended this

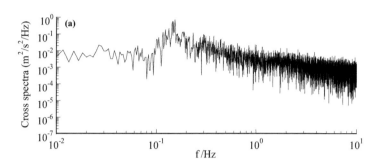

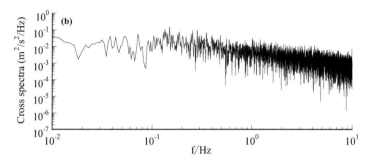

Figure 7.9 An example of energy spectrum density using a single sensor (a) and using two sensors (b). The u and w in (a) are from ADV #5. The u_1 and w in (b) are from ADV #5, and u_2 in (b) is from ADV #6 of trial 9. (Modified from Yuan et al. (2014).)

relationship and defined the turbulent kinetic energy (k) method by including three components of velocity to calculate the turbulence shear stress as:

$$\tau(z) = C_1 \rho k = C_1 \rho \left[0.5 \left(u'^2 + v'^2 + w'^2 \right) \right] \qquad (7.11)$$

where k is turbulent kinetic energy, $k = 0.5 \left(u'^2 + v'^2 + w'^2 \right)$ represents the fluctuations of the cross-stream velocity, and C_1 is a proportionality constant (0.19). This method has been applied to analyze shear stress in estuarine, coastal flow, shelf, and oceanography (Stapleton and Huntley 1995; Williams et al. 1999; Kim et al. 2000; Huthnance et al. 2002).

The turbulent kinetic energy (w') method is a modification of the turbulent kinetic energy (k) method (Kim et al. 2000). The turbulent kinetic energy (w') method uses only vertical velocity fluctuations as the instrument noise errors associated with vertical velocity variances are smaller than the noise errors for horizontal velocity fluctuations (Voulgaris and Trowbridge 1998; Biron et al. 2004). Thus, Equation (7.12) becomes

$$\tau(z) = C_2 \rho w'^2 \qquad (7.12)$$

where C_2 is the proportionality constant (0.9) (Kim et al. 2000).

Turbulent shear stress at P2 and P3 were estimated by Reynolds stress, the turbulent kinetic energy (k) method, and the turbulent kinetic energy (w') method, as described in Equations (7.10–7.12). The shear stress for the turbulent kinetic energy (k) method was predicted by calculating u'^2, v'^2, and w'^2. The shear stress for the turbulent kinetic energy (w') method was predicted by calculating w'. Turbulent intensities u'^2, v'^2, and w'^2 were calculated by var(u) for a single velocity sensor with the error produced by surface waves. For two velocity sensors, these were calculated by var(Δu), where Δu is the time series of the simultaneous difference of the streamwise velocity u_1 and u_2 recorded by two velocity sensors.

Figure 7.10 shows the dimensionless turbulent shear stress at P2 and P3 as a function of the relative freeboard using the three turbulent shear stress calculation methods.

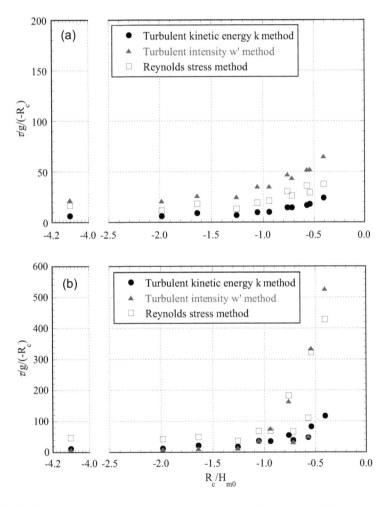

Figure 7.10 Dimensionless turbulent shear stress $\tau/g/(-R_c)$ calculated by turbulent kinetic energy k method, turbulent intensity w^- method, and Reynolds stress method: (a) P2 on the levee crest, and (b) P3 on the land-side levee slope. (Adapted from Yuan et al. (2014). Reproduced with permission from the Coastal Education and Research Foundation.)

The calculated shear stresses at P2 using the three methods were comparable, but the turbulent kinetic energy (w') method and Reynolds stress method overestimated the shear stress at P3. Nadal and Hughes (2009) indicated that the estimate of mean bottom shear stress on the land-side slope was approximately 220 Pa for −0.217 m freeboard and 0.534 m significant wave height, whereas the turbulent kinetic energy (w') method and Reynolds stress method overestimated the shear stress and had maximum shear stress of 1,200–1,500 Pa at P3. Because P3 is located on the land-side slope with high turbulence and shallow water, it was difficult to measure the 3D velocity, especially the vertical velocity (w), which plays an important role in turbulence methods for shear stress estimation, especially when applying the Reynolds stress and the turbulent kinetic energy (w') methods. However, the turbulent kinetic energy (k) method does not depend on the vertical velocity (w) alone which makes it more suitable for calculating the shear stress at P3.

7.4.3 Nadal and Hughes' method

Nadal and Hughes (2009) estimated the shear stress on the land-side slope of the levee during the nonuniform and unsteady overtopping flow using the one-dimensional momentum formula:

$$\tau_0 = \rho g \left(\frac{h_1 + h_2}{2} \right) \left(\sin\theta - \frac{h_2 - h_1}{s_{2,1}} - \left[\frac{u_2^2(I) - u_1^2(I)}{2gs_{2,1}} \right] - \left\{ \frac{\left[u_2(I) - u_2(I+1)\right] + \left[u_1(I) - u_1(I+1)\right]}{2g\left[t(I) - t(I+1)\right]} \right\} \right)$$

(7.13)

where h_1 and h_2 are flow thickness at the upstream and the downstream locations, respectively; θ is land-side slope angle; $s_{2,1}$ is the distance along the slope between upstream and the downstream locations; $u_1(I)$ and $u_2(I)$ are streamwise velocities at upstream and downstream locations, respectively; $u_1(I+1)$ and $u_2(I+1)$ are velocities at upstream and downstream locations one-time interval later, respectively; $t(I)$ = time at step I; and $t(I+1)$ = time at step $I+1$.

The streamwise velocity data of ADV5–ADV8 was used to estimate the shear stress of the area between ADV5 and ADV7 locations following the Nadal and Hughes' method. The upstream section velocity (u_1) is the average between the ADV5 and ADV6 velocities, and the downstream section velocity (u_2) is the average between the ADV7 and ADV8 velocities. The land-side slope angle was fixed at $\tan^{-1}(1/3)$, and the time history of the water depth (h) was collected using the acoustic range finders (Figure 7.3).

There are two other approximations of Equation (7.13). The first approximation, based on the steady-state flow assumptions, reduces the momentum formula to the weight of the water offset by the bottom shear stress, shown in Equation (7.14) as

$$\tau_{01} = \rho g \frac{(h_1 + h_2)}{2} \sin\theta$$

(7.14)

For the second approximation, Equation (7.14) is modified by incorporating the water depth change term between two points as in Equation (7.15) (Nadal and Hughes 2009).

Table 7.3 Turbulent shear stress estimates using Hughes and Nadal (2009) method

Trial number	R_c/H_{m0}	τ_0 (N/m²)	τ_{01} (N/m²)	τ_{02} (N/m²)
1	−1.64	660 ± 21	523 ± 5	650 ± 5
2	−1.057	654 ± 14	513 ± 7	637 ± 11
3	−0.762	639 ± 28	524 ± 18	589 ± 19
4	−0.544	570 ± 32	510 ± 45	550 ± 45
5	−0.406	525 ± 34	494 ± 62	501 ± 57
6	−4.102	714 ± 10	607 ± 8	744 ± 10
7	−1.983	699 ± 18	593 ± 8	736 ± 9
8	−1.261	694 ± 30	567 ± 20	683 ± 46
9	−0.941	723 ± 9	554 ± 15	675 ± 15
10	−0.719	644 ± 38	527 ± 55	620 ± 42
11	−0.573	604 ± 42	489 ± 70	536 ± 52

Note: Turbulent shear stress τ_0 is estimated with Equation (7.15), shear stress τ_{01} is estimated with Equation (7.16), and shear stress τ_{02} is estimated with Equation (7.17). The value after ± is the SD. After Yuan et al. (2014).

$$\tau_{02} = \rho g \left(\frac{h_1 + h_2}{2} \right) \left(\sin\theta - \frac{h_2 - h_1}{s_{2,1}} \right) \tag{7.15}$$

Table 7.3 lists the shear stress estimates using the Equations (7.13–7.15), and it shows that τ_0 and τ_{02} are very close, and τ_{01} is relatively small as the Equation (7.15) does not consider the water depth change from P2 to P3. Thus, the results of Nadal and Hughes' method were much larger than those calculated with the other three methods.

7.4.4 New formulas for shear stress estimation on the crest and on the land-side slope

Among the five shear stress estimation methods, the turbulent kinetic energy (w') method and the Reynolds stress method overestimated the shear stress on the land-side slope. The log profile method can only be used to calculate the shear velocity and shear stress at the crest. Nadal and Hughes' method cannot estimate the turbulent shear stress at an exact location. Therefore, the turbulent kinetic energy (k) method is the most suitable method to estimate the turbulent shear stress along the land-side slope during combined wave and surge overtopping.

Empirical formulas were established to relate the calculated mean shear stresses at P1 to the combined overtopping parameters. The best correlation was observed between the dimensionless mean shear stress to the relative freeboard (Figure 7.11a). Similar to the empirical formulas of turbulent kinetic energy (Equations 7.5 and 7.6), there is the same critical value, $-R_c/H_{m0}=2.0$, for the turbulent shear stress. Equation (7.16) gives the best-fit formula for the mean shear stress at P1. The formula shown by the solid curve in Figure 7.11a has a correlation coefficient of 0.904 and an RMS percent error of 0.407.

$$\frac{\tau_{P1}}{g(-R_c)} = \begin{cases} 4.323\left(-R_c/H_{m0}\right)^{-0.241}, & -R_c/H_{m0} < 2.0 \\ 3.77 & , \quad -R_c/H_{m0} > 2.0 \end{cases} \tag{7.16}$$

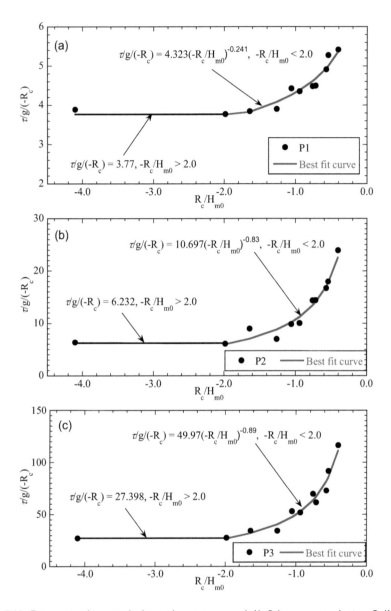

Figure 7.11 Dimensionless turbulent shear stress $\tau/g/(-R_c)$ versus relative R_c/H_{m0} at (a) P1, (b) P2, and (c) P3. (Adapted from Yuan et al. (2014). Reproduced with permission from the Coastal Education and Research Foundation.)

Equations (7.17) and (7.18) give the empirical formulas to estimate turbulent shear stress at P2 and P3. Figure 7.11b and c show the best correlations.

$$\frac{\tau_{P2}}{g(-R_c)} = \begin{cases} 10.697\left(-R_c/H_{m0}\right)^{-0.83}, & -R_c/H_{m0} < 2.0 \\ 6.232, & -R_c/H_{m0} > 2.0 \end{cases} \tag{7.17}$$

$$\frac{\tau_{P3}}{g(-R_c)} = \begin{cases} 49.97\left(-R_c/H_{m0}\right)^{-0.89}, & -R_c/H_{m0} < 2.0 \\ 27.398, & -R_c/H_{m0} > 2.0 \end{cases} \tag{7.18}$$

Application of Equations (7.16–7.18) is limited to the range of tested hydraulic parameters and laboratory experimental conditions, such as the seaward-side slope of 1V:4.25H, and the land-side slope of 1V:3H.

Chapter 8

Hydraulic erosion on landward-side slope of levees and conceptual model of soil loss from levee surface

This chapter introduces two parts. The first part is the analyses of hydraulic erosion of the landward-side slope of levees. The erosion data of three strengthening systems measured during full-scale flume test are introduced and their erosion characteristics and resistance are analyzed. The second part is the conceptual model of the erosion of levees. Based on the data of full-scale flume tests and erosion function apparatus (EFA) tests, a conceptual model for the overtopping erosion on the landward-side slope of levees and the process of levee failure are provided to describe the reason for the differences between the existing research on the erosion and failure of levees. In addition, failure modes and testing methods of HPTRM are explained.

8.1 Hydraulic erosion on landward-side slope of levees

Based on erosion measurements during full-scale flume tests, the hydraulic erosion of RCC, ACB, and HPTRM under combined wave and surge overtopping conditions are analyzed.

8.1.1 Hydraulic erosion on the RCC test section

The initial thickness of RCC was 30 cm in the RCC test section with a relatively rough and loose top layer. During combined wave and surge overtopping, much of this loose material was washed away freely, leaving various patches of lower layers exposed. The damage was limited to surface erosion or polishing where the loss of surface materials ranged from none to 25 mm or less. During the RCC tests, according to the observed surface hydraulic erosion, the area observation of the overall erosion status and the erosion depth of the local erosion points were performed. In this section, the hydraulic erosion of RCC during combined wave and surge overtopping are analyzed based on the changes in the erosion area and erosion depth.

8.1.1.1 General erosion status

To understand the erosion status of the RCC surface, three grades of hydraulic erosion were defined: deep erosion, shallow erosion, and no erosion (Figure 8.1). Deep erosion means that a significant hole appears on the RCC surface. Shallow erosion falls into two categories: (1) shallow erosion with gravel exposure (gravel exposure, but no significant hole formed); and (2) shallow erosion with cement surface (no gravel exposure,

Figure 8.1 Three grades of hydraulic erosion on the RCC test section: (a) deep erosion, (b) shallow erosion with gravel exposure, (c) shallow erosion with
cement surface, and (d) no erosion. (Adapted from Li et al. (2012). Reproduced with permission from Elsevier.)

but a thin layer of cement has eroded). No erosion indicates that no significant cement
is washed away from the RCC surface.

During the entire testing period of combined wave and surge overtopping tests,
four erosion checks were done for the preselected four areas on the RCC test section
(Figure 8.2). The first area was located on the crest of the levee. The second area was
located on the top of the landward-side slope. The third area was located on the upper
part of the landward-side slope. The last area was located in the middle of the landward-side slope. Four erosion checks included the following:

1   The first check was conducted after six test trials in surge-only overflow and four
    test trials in combined wave and surge overtopping.
2   The second check was done after the seventh test trial in combined wave and surge
    overtopping.
3   The third check was conducted after the ninth test trial in combined wave and
    surge overtopping.
4   The last check was done after all tests were completed.

In the erosion check, the surface of RCC was measured using a hollow frame of 30.48
cm×30.48 cm (1 ft×1 ft) (Figure 8.3). The inner area of the frame was divided into

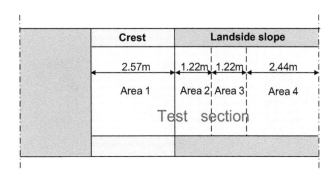

Figure 8.2 Preselected areas for erosion inspection. (Adapted from Li et al. (2012).
Reproduced with permission from Elsevier.)

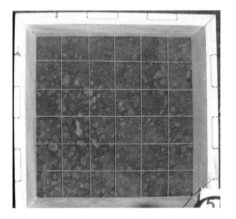

Figure 8.3 Observation of surface erosion of RCC: grids (a) and observation process (b).

6×6 grids with equal spacing by fine lines. After being placed on the surface of RCC, the erosion status (deep erosion, shallow erosion, or no erosion) inside each grid was recorded. At the end of the measurements, the percentage of three grades of hydraulic erosion grid areas in the total area of the region was counted in each region.

Figures 8.4–8.7 present the percentage of three types of surface erosion in the four preselected areas. For all the preselected areas during the entire testing period, more than 70% of the areas indicated no erosion. Most erosion occurs before the first check, and between every two checks, the size of the erosion areas increases slightly (almost negligible). This may be explained by the fact that erosion usually occurs due to the initial washing away of the loose materials on the RCC surface and occurs at places where the surface material is loose after the first test. Hence, if the surface is not eroded after the first test, it can be concluded that further erosion will be very difficult to observe unless much greater hydraulic force is present. The erosion of the interface zone between no erosion area and erosion area accounts for the very little increase in the erosion area. Although the total erosion area does not increase much, the shallow erosion may turn into deep erosion, which is significant in area 3, as shown in Figure 8.6. Area 2 and area 3 had relatively more erosion percentage because they were located on the top of the land-side slope, where there is the strongest combined supercritical flow and dynamic impact of wave attack during the tests.

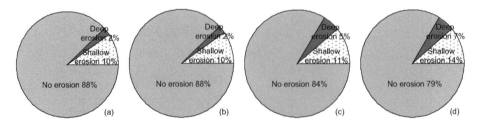

Figure 8.4 Hydraulic erosion inspection on the RCC test section in the area (i) of (a) check 1, (b) check 2, (c) check 3, and (d) check 4. (Adapted from Li et al. (2012). Reproduced with permission from Elsevier.)

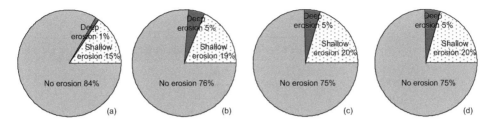

Figure 8.5 Hydraulic erosion inspection on the RCC test section in the area (ii) of (a) check 1, (b) check 2, (c) check 3, and (d) check 4. (Adapted from Li et al. (2012). Reproduced with permission from Elsevier.)

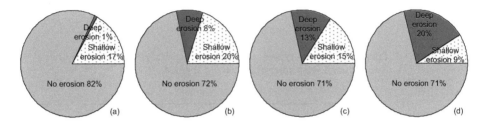

Figure 8.6 Hydraulic erosion inspection on the RCC test section in the area (iii) of (a) check 1, (b) check 2, (c) check 3, and (d) check 4. (Adapted from Li et al. (2012). Reproduced with permission from Elsevier.)

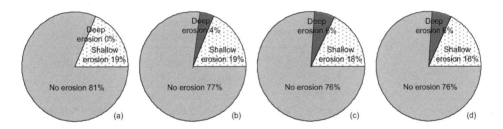

Figure 8.7 Hydraulic erosion inspection on the RCC test section in the area (iv) of (a) check 1, (b) check 2, (c) check 3, and (d) check 4. (Adapted from Li et al. (2012). Reproduced with permission from Elsevier.)

8.1.1.2 Depth monitoring of four typical deep erosion spots

After the test trial 1 in surge-only overtopping tests, four typical deep erosion spots were chosen to monitor the erosion depth (E1, E2, E3, E4). The locations of the deep erosion spots are shown in Figure 8.8. A Vernier caliper was used to measure the changes in depths of each spot. A ruler was used to set the no erosion level of the RCC surface.

The histories of the erosion depths at the deep erosion spots are illustrated in Figure 8.9. Figure 8.9 shows that most of the erosion occurs during the first test (surge overflow trial 1). During the entire testing period, the erosion depths for all four spots

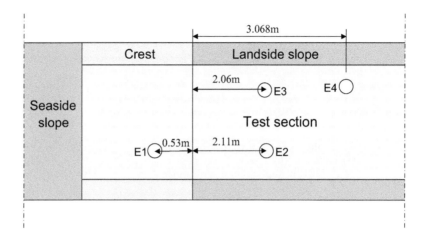

Figure 8.8 Locations of four typical deep erosion spots. (Adapted from Li et al. (2012). Reproduced with permission from Elsevier.)

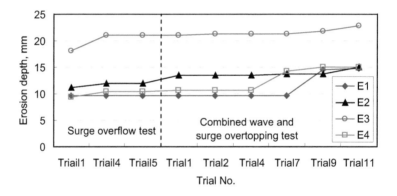

Figure 8.9 Erosion depth histories of four typical deep erosion spots. (Adapted from Li et al. (2012). Reproduced with permission from Elsevier.)

remained almost unchanged. This indicates that erosion is caused by the initial washing away of loose materials on the RCC surface, and that most of the erosion depth occurs during the first test. After the loose material at the surface was eroded, the erosion depth would only increase slightly at the same level of hydraulic power. It was noticeable that the erosion depths at all the deep erosion spots increased slightly during the last few tests. This indicated that the surface of RCC at the deep erosion spots may have become looser and erodible again after long-time washing and soaking.

8.1.2 Hydraulic erosion on the ACB test section

The physical model of the ACB mat is shown in Figure 8.10. There are 120 individual units and 30 one-and-half individual units in the ACB mat distributed in 30 rows. Each row of units was laterally offset by one-half of a block width from the adjacent row so

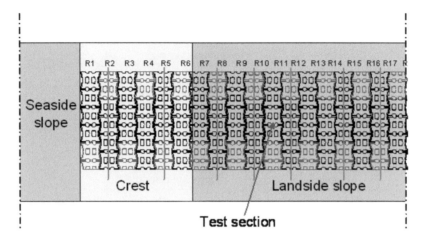

Figure 8.10 Selected locations for ACB block uplift and settlement inspection on the ACB test section.

that any given block could be cabled to four other blocks (two in the row above and two in the row below).

As shown in Figure 8.10, the erosion checks for the ACB test section included the following seven rows along the flow direction: R2, R5, R8, R10, R12, R15, and R17. There were five blocks in each row. In the middle of each ACB block, a laser level was used to measure the initial vertical elevation. After every trial, the vertical movement (uplift or settlement) of the top of the ACB blocks was checked (Figure 8.11).

The vertical elevation measurement results of ACB during the test period are shown in Figure 8.12. Some ACB blocks randomly lifted and settled because of the flow turbulence. The vertical movement was within ±5 mm. The different variations of uplift and settlement of ACB blocks could be related to the erosion and swelling of the

Figure 8.11 Survey of the ACB slope protection section in the selected location (a) and laser level (b).

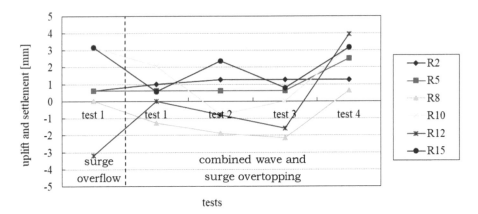

Figure 8.12 Vertical movement inspection on the ACB test section.

underlying clay soil. However, the vertical movement of ACB blocks was very small and the regularity was not obvious. Thus, there was no catastrophic failure in the ACB test section.

8.1.3 Hydraulic erosion on the HPTRM test section

The hydraulic erosion of the HPTRM test section is analyzed from three aspects, including experimental phenomena, soil loss, and grass stem and blade loss.

8.1.3.1 Experimental phenomena

Unlike RCC and ACB test sections, the water in the flume looked turbid and there were soil particles deposited on the top of the sea-side slope of the tested model (water in the flume was recycled, soil particles eroded through the circulation system) during the HPTRM test periods. The above observations showed that a certain amount of soil was carried away by water flow from HPTRM (Figure 8.13).

After all tests were completed, the overall condition of the HPTRM system was good with no major failure (Figure 8.14). The density of grass stems was still high. At some points, the HPTRM emerged slightly from the soil and the depth of exposure was about 0.5 cm (in Figure 8.15, the grass was brushed away by hand so that the soil surface could be seen).

8.1.3.2 Soil loss

After each test, soil surface elevations were measured at 64 positions on the crest and landward-side slope, as shown in Figure 8.16. The average values of the eight measurement points around P1–P5 were taken as the soil loss at each point. The measurement points in rows S11 and S12 were always submerged in the test intervals during test days. Consequently, the soils at S11 and S12 were looser and their erodibility was higher than the ones at the upper position. In soil surface elevation measurement, a special bench was used to determine the measurement datum elevation. The feet of the bench was

Figure 8.13 Soil particles deposited on the top of the sea-side slope of the tested model.

Figure 8.14 Observation of grass on the crest and landward-side slope after all over-topping tests in the full-scale levee embankment. (Adapted from Pan et al. (2013b). Reproduced with permission from the Coastal Education and Research Foundation.)

Figure 8.15 Observation of stems and exposure of HPTRM after all overtopping tests in the full-scale levee. (Adapted from Pan et al. (2013). Reproduced with permission from the Coastal Education and Research Foundation.)

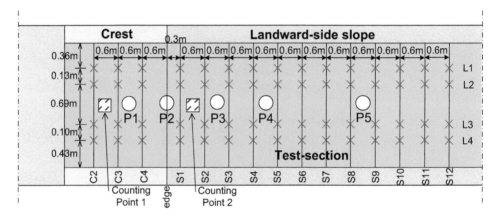

Figure 8.16 Soil loss monitoring points and grass counting positions on the crest and landward-side slope. (Adapted from Pan et al. (2013). Reproduced with permission from the Coastal Education and Research Foundation.)

placed on the cement edges of both sides of the test area and slid freely along the top of the levee and the landward-side slope. The distance between the soil surface and the top of the bench was measured with a steel ruler. The measurement process is shown in Figure 8.17.

Figure 8.18 shows the total soil loss measured after each overtopping test. After trial 4 tests, the soil losses from P1 to P5 did not increase significantly. It appeared that the maximum soil loss had been reached by then. The soil loss of the submerged part kept

Figure 8.17 The soil surface elevation measurement process.

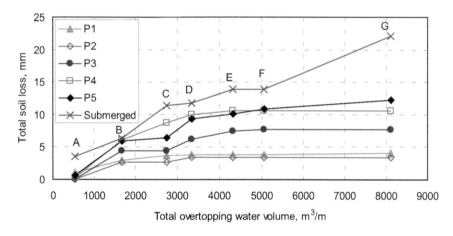

Figure 8.18 Total soil losses measured after each combined wave and surge overtopping test (Adapted from Pan et al. (2013). Reproduced with permission from the Coastal Education and Research Foundation.)

increasing during all tests because of the loose condition of the soil. This phenomenon is called the "upper limit" in this book.

Figure 8.19 illustrates this phenomenon. Before conducting any test, there is no void space between the HPTRM and the compacted clay layer underneath. The grass root penetrated through the HPTRM and clay and interlocked with the HPTRM (Figure 8.19a). During the tests, some soil in the clay layer underneath the HPTRM was eroded and a certain height of void space between the HPTRM and clay layer was

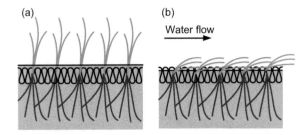

Figure 8.19 The characteristics of the HPTRM before and after the tests: (a) the profile of HPTRM before all the tests; (b) the profile of HPTRM after soil loss stops increasing significantly. (Adapted from Pan et al. (2013). Reproduced with permission from the Coastal Education and Research Foundation.)

formed (Figure 8.19b). The soil at each position was resistant to a certain range of hydraulic power because of the protection of the grass and HPTRM.

Figure 8.18 shows that the maximum soil losses increase along the landward-side slope. There are many factors affecting soil losses of slope protection, and HPTRM is more complex because of its vegetation and geogrid. Therefore, it was difficult to provide an effective estimation method for the relationship between maximum soil loss ("upper limit") and the highest average overtopping flow velocity with the limited test groups in this study. Based on the limited measured data, the estimation relationship based on the method of the "upper limit" can be expressed as follows:

$$E_{\max} = 11.23v_{ws} - 16.24 \tag{8.1}$$

where $E_{\max}$ is the maximum soil loss depth in mm, and v_{ws} is the average overtopping flow velocity in m/s which is calculated using average overtopping discharge and average flow thickness. However, Equation (8.1) is only used for research discussion and is not recommended for practical calculation at this point.

Although maximum erosion depth exists for certain ranges of hydraulic conditions, there is concern about how the erosion occurs before the maximum erosion depth is reached (Hughes 2008). Because the maximum erosion was almost reached after trial 4 in this study, the erosion rates of test trials 1–4 at P1, P3, P4, and P5 versus average overtopping velocities are shown in Figure 8.20. Assuming erosion increases linearly over time, the average erosion rate for each test was calculated by dividing the measured erosion depth by the testing duration. As shown in Figure 8.20, the erosion rate points are distributed relatively concentrated around the best-fit linear relationship, which is expressed as:

$$r_E = 5.3v_{ws} - 9.3 \tag{8.2}$$

where E is the erosion rate in mm/h. Figure 8.20 shows that erosion started to occur when the average overtopping flow velocity exceeded 1.75 m/s. The average overtopping flow velocity of 1.75 m/s was a threshold of soil loss from HPTRM. Beyond this

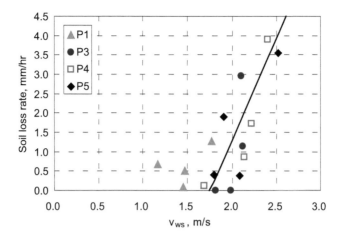

Figure 8.20 Erosion rate versus average overtopping velocity of each test at four positions: P1, P3, P4, and P5. The P1 is located in the middle of the levee crest. The P3, P4, and P5 are located in the land-side slope of the levee embankment.

threshold, the relationship between erosion rate and average overtopping flow velocity was approximately linear. Due to the presence of scattered data around the lower end of the best-fit line (Figure 8.20), Equation (8.20) only indicates the general distribution but does not provide an exact estimate for the erosion rates.

8.1.3.3 Grass stem and blade loss

After each overtopping test, the number of grass stems and blades were counted within a 76.2 mm by 76.2 mm (3 inch by 3 inch) wooden square at two locations (Figure 8.16). One of the locations was at the top of the crest between survey stations C2 and C3, and another one was on the upper portion of the land-side slope between survey stations S1 and S2. Grass counts were recorded by counting the number of stems within each square and the number of blades coming off of each stem. Figure 8.21 shows the Bermuda grass. The counting process is shown in Figure 8.22.

As shown in Figure 8.23, the stem densities remained almost unchanged at both counting locations with only negligible decreases throughout the experimental trials. Thus, within the range of tested hydraulic parameters, the stems were seldom taken away from the HPTRM by overtopping flow. During the first few tests, the densities of the blade kept decreasing. However, the densities of the blade at both counting locations remained unchanged after trial 4 (Figure 8.24).

The initial number of blades on each stem was about eight to nine (Figure 8.25). After trial 4, this number decreased to a constant value of about 3–4. Within the range of tested hydraulic parameters, there were three or four blades on average on each stem that were not washed away no matter how long the overtopping process lasted. Another finding was that the final extent of the decrease in the average number of

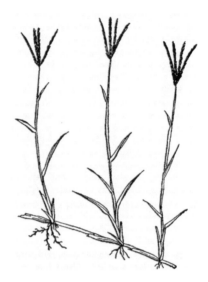

Figure 8.21 The Bermuda grass.

Figure 8.22 The counting process of grass stems and blades.

blades on each stem was the same for the two counting locations. However, on the upper portion of the landward-side slope at counting point 2, the decrease was faster. A possible reason for this observation could be that the shear stress was the main cause of blades leaving off, and it was larger at counting point 2 as it was at the top of the land-side slope.

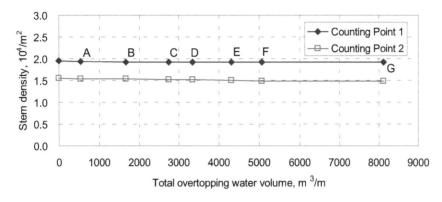

Figure 8.23 Observed grass stem density after each combined wave and surge over-topping test: (A) trial 1, (B) trial 2, (C) trial 3, (D) trial 4, (E) trial 5, (F) trial 6, and (G) trial 9. (Adapted from Pan et al. 2013. Reproduced with permission from the Coastal Education and Research Foundation.)

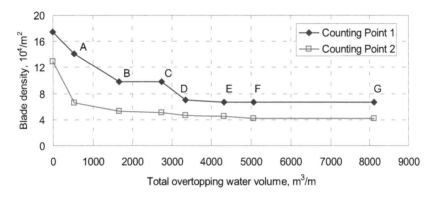

Figure 8.24 Observed blade density after each combined wave and surge overtopping test: (A) trial 1, (B) trial 2, (C) trial 3, (D) trial 4, (E) trial 5, (F) trial 6, and (G) trial 9. (Adapted from Pan et al. (2013). Reproduced with permission from the Coastal Education and Research Foundation.)

8.2 Conceptual model of soil loss from levee surface

8.2.1 Definition

Two issues exist in the domain of the erosion and failure process of the earthen dam and levees. The first is the choice of the hydraulic parameter that is used to evaluate the erodibility of the dam/levee, and the issue is the position of the erosion starting spot.

8.2.1.1 Choice of the hydraulic parameter

Usually, both flow velocity and shear stress have been used in the evaluation of the erodibility. Nelson (2005) conducted a study on the permissible shear stress and flow

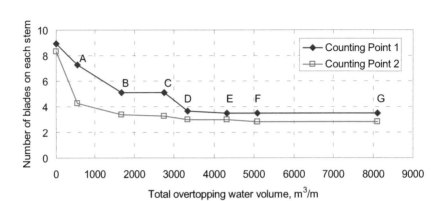

Figure 8.25 Observed number of blades on each stem after each combined wave and surge overtopping test: (A) trial 1, (B) trial 2, (C) trial 3, (D) trial 4, (E) trial 5, (F) trial 6, and (G) trial 9. (Adapted from Pan et al. (2013). Reproduced with permission from the Coastal Education and Research Foundation.)

velocity of TRM based on a surge-only overflow experiment. Briaud (2008) indicated that the erodibility can be defined as the relationship between the velocity/shear stress of the water flowing over the soil and the corresponding erosion rate experienced by the soil. Many scholars have used both flow velocity and shear stress as the key parameters to determine the erodibility while some used only one.

For Nelson's study of open-channel steady overflow and Briaud's study of constant pressure flow, the divergence of key parameters is not important as the flow velocity and shear stress are one-to-one correlations. In combined wave and surge overtopping, the distributions of flow velocity and shear stress are not the same as the calculation of shear stress involves the partial derivative of flow velocity to time, the partial derivative of flow velocity and flow thickness to space. The question is, at least in the case of wave overtopping and landward-side slope erosion, which one is the key factor, shear stress or flow velocity, that controls the soil loss, or which one should be used in the evaluation of the soil erodibility? To determine whether flow velocity or shear stress should be adapted in the evaluation of the erodibility, the average shear stresses and average overtopping flow velocities at P1, P3, P4, and P5 were plotted in Figure 8.26. Because an "upper limit" exists for erosion, only the data before reaching the "upper limit" for each point was plotted. The correlation between the shear stress and the erosion rate is not clear while a relationship can be found between the flow velocity and the erosion rate (Figure 8.26). A threshold velocity is noted after which erosion starts to occur. The values of the threshold velocities ranged from 1.5 to 2.0 m/s for different points. After the threshold velocity was reached, the relationship between erosion rate and average flow velocity was approximately linear. Therefore, it was concluded that the flow velocity was better correlated to the soil erosion along the levee crest and landward-side slope in the vegetated HPTRM system case. It should be mentioned that the use of the threshold velocity should be limited to the overtopping condition and the linear relationship should be limited to the tested range. As a result of this finding, it should be explained why flow velocity and soil erosion have a better correlation, and it

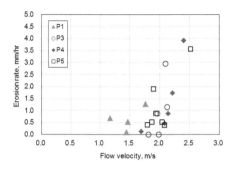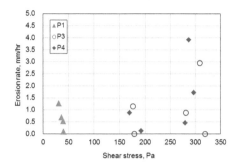

Figure 8.26 Erosion rate versus average shear stress (a) and average flow velocity (b) at the monitoring points. (Adapted from Pan et al. (2015b). Reproduced with permission from Elsevier.)

should be checked whether this conclusion can be extended to other conditions. This has become a question to be discussed by the conceptual model of soil loss from the levee surface.

8.2.1.2 Position of erosion starting spot

Many researchers have different approaches to the position of erosion starting spot of levees. According to existing studies, the erosion starting spot could be the top of the dam (e.g., Johnson and Illes 1976; MacDonald and Langridge-Monopolis 1984), landward-side slope (e.g., Hahn et al. 2000), or the toe of the dam (e.g., Ralston 1987; Powledge et al. 1989). Figure 8.18 shows that the maximum soil losses increase along the landward-side slope. Although there was no obvious damage, it was not difficult to infer that the surface damage process should have begun at the toe of the dam. Therefore, how to explain the different initial erosion positions in different studies becomes another question for the conceptual model of soil loss from the levee surface.

8.2.2 Soil erodibility of HPTRM

Based on the analysis of hydraulic parameters and the position of the erosion starting spot, two erosion characteristics of vegetated HPTRM system under combined and surge overtopping are defined. The first is the hydraulic parameter which is the average flow velocity on the landward-side slope, and the second is the erosion starting spot which could be the toe of the dam.

As the analysis in the previous section showed that the flow velocity was better correlated to the soil erosion along the levee crest and landward-side slope in vegetated HPTRM system case, the results of EFA and the the large-flume test were drawn together in the erodibility classification chart from Briaud et al. (2008). The EFA test results of the samples are shown in the erodibility classification charts of velocities in Figure 8.27. As shown in the erodibility classification charts of velocities, the clay sides of the samples mainly fall in Category II (high erodibility), while the grass sides

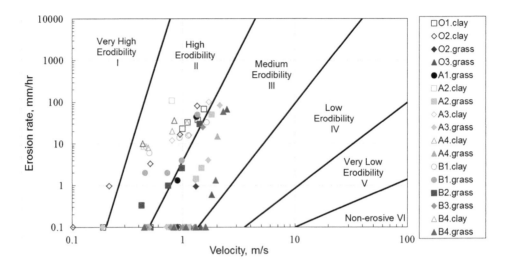

Figure 8.27 EFA test results plotted in the erodibility classification charts of velocities. (The erodibility classification charts are from Briaud et al. (2008). Adapted from Pan et al. (2015). Reproduced with permission from Elsevier.)

of the samples mainly fall in Category III (medium erodibility). The existence of turf reduces soil erodibility from high to medium erodibility. According to the erodibility classification chart, HPTRM system plays the same role. However, the data point of the flume test shown in Figure 8.27 is the data before reaching the "upper limit". Therefore, combined with the analysis of Section 7.1.3, it can be concluded that the erodibility of the overlying layer of HPTRM is close to that of ordinary turf, but the existence of the "upper limit" is one of the characteristics that it is better than ordinary turf.

8.2.3 Conceptual model of soil loss from levee surface

It is assumed that two types of levee erosion mechanisms (i.e., particle erosion and block erosion) correspond for the levee erosion on the crest and landward-side slope. Particle erosion and block erosion are related to microscopic shear force and macroscopic shear force, respectively. The microscopic shear force indicating the sediment carrying force acting on single particles is directly related to the flow velocity. The macroscopic shear force indicating the wave/flow action on the levee surface is induced by the act of the wave and the change in flow velocity, and can be calculated using the Saint-Venant equations (Sturm 2001; Nadal and Hughes 2009) or the velocity profile (Gudavalli et al. 1997). It is also easy to test the shear strength (threshold shear stress) of the levee surface material. The particle erosion is induced by the microscopic shear force. In particle erosion (Figure 8.28a), soil particles are taken away from the levee surface one by one by the flow. As a result, particle erosion increases gradually and smoothly and the degree of erosion is directly affected by the microscopic shear force, the weight of the particle, and the forces among particles. After adequate particles

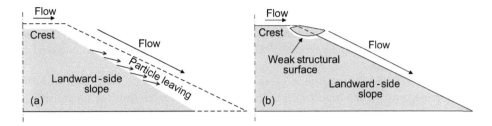

Figure 8.28 Two types of erosion mechanisms of soil loss from levee surface: (a) particle erosion and (b) block erosion. (Adapted from Pan et al. (2015b). Reproduced with permission from Elsevier.)

have left the surface, a rill occurs on the levee surface. Block erosion is induced by the macroscopic shear force. When the macroscopic shear stress reaches the shear strength of the weakest structural surface near the levee surface, a small block of soil is taken away from the levee surface and erosion starts to occur during block erosion (Figure 8.28b). As the shear stress increases and the exposure of new weak structure surfaces occurs, the levee surface is destroyed at the weak structure surfaces and soil is taken away from levee surface block by block. If shear stress keeps increasing, a huge amount of soil would be taken away and a breach would occur. Therefore, block erosion occurs at the maximum shear stress in a relatively short process.

Based on the hypothesis of the two different erosion mechanisms mentioned above, the observations that the erosion is triggered at different locations of the dam/embankment can be explained. In an overtopping case, both types of erosion may occur. The failure of the levee caused by the particle erosion could be defined by an erosion depth which is large enough to form a rill on the levee surface. The failure of the levee caused by block erosion could be defined by a threshold shear stress which causes the weakest structure surfaces starting to be destroyed. Different types of levees/embankments have different erodibilities against microscopic shear force and different macroscopic shear strengths. Therefore, different erosion starting spot can be observed in different cases.

In the HPTRM overtopping case, the shear strength of the levee surface was enhanced largely by HPTRM, and as a result, the particle erosion occurred first. This mechanism was the reason why the erosion was related to the flow velocity rather than the shear stress in the HPTRM-strengthened case. Because particle erosion was more related to flow velocity, the erosion became larger in a lower position with a large average flow velocity. It should be noted that because of the protective effect of HPTRM discussed earlier, the flow velocity-induced erosion stopped when the "upper limit" of soil loss was reached.

As the hypothesis of the two types of erosion mechanisms explained the flume test data and EFA test data well and answered the shear stress or flow velocity question and the erosion starting point question, it was considered that the hypothesis was reasonable. However, because of the limited sample size, the conceptual model was qualitative and given tentatively. Additional test data is needed to arrive at more quantitative conclusions (Yuan et al. 2015).

8.2.4 Failure modes of HPTRM-strengthened levee

As HPTRM raises the erosion resistance of the vegetation lining largely, it is difficult to model the failure process of the vegetated HPTRM system in the lab. In this section, the failure process of the vegetated HPTRM system is given tentatively based on the conceptual model.

For the HPTRM-strengthened levee, the HPTRM mat prevents block erosion by subjecting most of the macroscopic shear stress and restricting the movement of soil blocks. Therefore, the failure induced by macroscopic shear stress can only be caused by the destruction of the HPTRM mat itself or the tearing off of the HPTRM mat from the levee surface. Noticing that the strength (ultimate tensile strength=34.2 kN/m, see Section 3.1.2) of the HPTRM mat is much higher than the adhesion strength between the HPTRM mat and the levee surface, the mode of the failure induced by the macroscopic shear stress is the tearing off of the HPTRM mat from the levee surface (Figure 8.29a). When water flow is fast enough, the transient shear stress induced by a large wave may exceed the adhesion strength between the HPTRM mat and the levee surface at one weak position, leading to the tearing off the HPTRM mat, and then failure occurs. If water flow cannot tear the HPTRM mat off from the levee surface in a short time, the HPTRM-strengthened levee would suffer from continuous particle erosion and may also eventually fail. In this case, the occurrence of failure requires two conditions, that is, flow intensity and duration. First, the water flow should be powerful enough so that the "upper limit" of soil erosion is higher than the thickness of the HPTRM mat. Second, the overtopping duration needs to be long enough to allow erosion to develop at the bottom of the HPTRM mat (Figure 8.29b). Without the adhesion force between the HPTRM mat and the levee surface, the HPTRM mat would easily be torn off causing failure. However, in an overtopping case in which the water flow is not able to tear the HPTRM mat off at first (the same as the beginning case (b) (Figure 8.29)), the failure may occur before the erosion develops to the bottom of the HPTRM mat. This can happen because of the adhesion strength between the HPTRM mat and the levee surface would keep decreasing in the process of the development of particle erosion by thinning of the soil in the HPTRM mat. When the adhesion strength reaches the transient shear stress induced by the largest waves, the

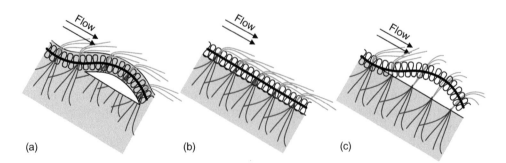

Figure 8.29 Failure modes of HPTRM-strengthened levee: (a) torn off of the HPTRM mat; (b) erosion develops to the bottom of the HPTRM mat; (c) torn off of the HPTRM mat in the process of the development of particle erosion. (Adapted from Pan et al. (2015). Reproduced with permission from Elsevier.)

HPTRM mat is torn off and failure happens (Figure 8.29c). As the adhesion strength is usually large, the combined wave and surge overtopping cases that produce much larger instantaneous shear stress are more dangerous than steady surge-only overflow cases.

According to the above failure modes of HPTRM-strengthened levee, the parameters related to the safety of the landward-side slope of HPTRM-strengthened levee included critical erosion velocity, the relationship between erosion rate and flow velocity, the relationship between "upper limit" and flow velocity, and bonding force between geogrid and HPTRM. Critical erosion velocity is mainly affected by soil properties. It can be measured by EFA tests or small flume tests. Erosion rate and "upper limit" are related to flow velocity. A series of small steady flow laboratory tests with different flow velocities can be used to measure the erosion rate. It is noted that the test time should be long enough to ensure that the erosion amount reaches the "upper limit". The bonding strength between geogrid and HPTRM is mainly affected by the thickness of soil thickness in geogrid. Therefore, pull-out tests of geogrid can be conducted for samples with different thicknesses of the soil layer in geogrid to obtain the results of geogrid and HPTRM at different particle erosion levels (Figure 8.30). Based on this test method, the test of HPTRM can be conducted in a small laboratory test facility without repeating large-scale flume tests which could be costly (Pan et al. 2018).

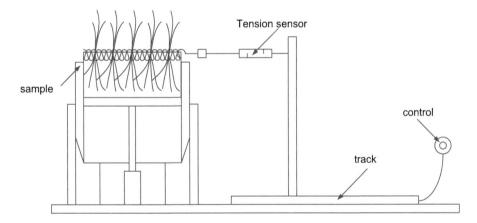

Figure 8.30 Pull-out tests of HPTRM.

Chapter 9

Numerical study of combined wave overtopping and storm surge overflow of strengthened levee

Hydrodynamics at the toe of the landward-side levee slope is the concern of many studies where the flow has high speed, high turbulence, and small thickness. It is also the location of the largest erosion. However, limitations of physical scale or experimental equipment make it difficult to understand the supercritical overtopping flow at the toe of the landward-side slope. Numerical models are often employed to fill the gaps that physical models have because of the limitations of instruments in the simulation of wave overtopping. However, there is very limited information about the effects of combined wave overtopping and storm surge overflow on the strengthened levee, especially at the toe of the landward-side slope. This chapter presents the two numerical methods that are used to investigate the overtopping hydraulics at the toe of the landward-side slope of HPTRM- and RCC-strengthened levee under combined wave overtopping and storm surge overflow.

9.1 Princeton Ocean Model (POM) method

POM is a widely used ocean model used to simulate circulation and mixing processes in rivers, estuaries, shelf and slope, lakes, semi-enclosed seas, and open and global ocean (Blumberg and Mellor 1987; Ezer et al. 2003; Ezer et al. 2008; Li et al. 2015). POM is a sigma coordinate, free surface ocean model with embedded turbulence and wave submodels. A wetting and drying (WAD) scheme has been implemented into the POM (named POM-WAD) to simulate flow in near-coast regions where WAD processes prevail (Oey 2005).

In this chapter, the POM-WAD model was used to simulate the incidence of random waves on an HPTRM strengthened levee during the combined wave overtopping and storm surge turbulent overflow. The model was verified with empirical equations and experimental results for storm surge overflow and full-scale combined wave overtopping and storm surge overflow. After the calibration, the model was used to investigate the hydraulic phenomena of combined wave overtopping and storm surge overflow in the HPTRM strengthened levee, especially at the toe of the landward-side slope. The erosion on the HPTRM strengthened levee was not within the scope of this study. Therefore, it was not discussed in this book.

9.1.1 Numerical methodology

9.1.1.1 Governing equations

The Navier–Stokes equations under the assumption of hydrostatic pressure and Bouss-inesq approximation are used in POM-WAD for continuity equation and momentum equations (Li et al. 2015a):

$$\frac{\partial u}{\partial x}+\frac{\partial v}{\partial y}+\frac{\partial w}{\partial z}=0 \tag{9.1}$$

$$\frac{\partial u}{\partial t}+u\frac{\partial u}{\partial x}+v\frac{\partial u}{\partial y}+w\frac{\partial u}{\partial z}=-g\frac{\partial \zeta}{\partial x}+\frac{\partial}{\partial x}\left(\varepsilon_h\frac{\partial u}{\partial x}\right)+\frac{\partial}{\partial y}\left(\varepsilon_h\frac{\partial u}{\partial y}\right)+\frac{\partial}{\partial z}\left(\varepsilon_z\frac{\partial u}{\partial z}\right) \tag{9.2}$$

$$\frac{\partial v}{\partial t}+u\frac{\partial v}{\partial x}+v\frac{\partial v}{\partial y}+w\frac{\partial v}{\partial z}=-g\frac{\partial \zeta}{\partial y}+\frac{\partial}{\partial x}\left(\varepsilon_h\frac{\partial v}{\partial x}\right)+\frac{\partial}{\partial y}\left(\varepsilon_h\frac{\partial v}{\partial y}\right)+\frac{\partial}{\partial z}\left(\varepsilon_z\frac{\partial v}{\partial z}\right) \tag{9.3}$$

$$\frac{\partial P}{\partial z}=-\rho g \tag{9.4}$$

where t is time, u, v, and w are the flow velocity components in the x, y, and z directions, respectively, ζ is the sea level, z is the vertical coordinate increasing upward with $z=0$ located at the undisturbed water surface and positive upward, P is the water pressure, ρ is the water density, g is the gravitational acceleration, and ε_h and ε_z are the eddy viscosity of turbulent flow in the horizontal and vertical directions, respectively.

The Mellor–Yamada Level 2.5 turbulence closure model (Mellor and Yamada 1982) and a prognostic equation (Mellor et al. 1998) for the turbulence macroscale are used to calculate the vertical eddy viscosity and diffusivity. The Mellor–Yamada Level 2.5 turbulence closure includes two partial differential equations to compute the turbu-lent kinetic energy (q^2) and a turbulent macroscale (l). The equation for the turbulent kinetic energy (without considering the variation of flow density) is:

$$\frac{\partial q^2}{\partial t}+u\frac{\partial q^2}{\partial x}+v\frac{\partial q^2}{\partial y}+w\frac{\partial q^2}{\partial z}=\frac{\partial}{\partial x}\left(\varepsilon_q\frac{\partial q^2}{\partial x}\right)+\frac{\partial}{\partial y}\left(\varepsilon_q\frac{\partial q^2}{\partial y}\right)+\frac{\partial}{\partial z}\left(\varepsilon_q\frac{\partial q^2}{\partial z}\right)+2\left(P_s-\frac{q^3}{B_1 l}\right) \tag{9.5}$$

and the equation for the turbulent macroscale is:

$$\frac{\partial q^2 l}{\partial t}+u\frac{\partial q^2 l}{\partial x}+v\frac{\partial q^2 l}{\partial y}+w\frac{\partial q^2 l}{\partial z}=\frac{\partial}{\partial x}\left(\varepsilon_q\frac{\partial q^2 l}{\partial x}\right)+\frac{\partial}{\partial y}\left(\varepsilon_q\frac{\partial q^2 l}{\partial y}\right)+\frac{\partial}{\partial z}\left(\varepsilon_q\frac{\partial q^2 l}{\partial z}\right)$$
$$+lE_1 P_s-\frac{q^3}{B_1}\left[1+E_2\left(\frac{l}{\kappa L}\right)^2\right] \tag{9.6}$$

where P_s is the shear production, defined as $P_s=\varepsilon_z\left(\frac{\partial u}{\partial z}\right)^2+\varepsilon_z\left(\frac{\partial v}{\partial z}\right)^2$, $q^3/B_1 l$ is the tur-bulent dissipation, L is defined as $(\zeta-z)^{-1}+(H'+z)^{-1}$, H' is the average water depth at mean water level, $\varepsilon_q(=q l/\chi_q)$ is the eddy diffusion coefficient for turbulence energy, ε_z is the vertical eddy viscosity, constant E_1 is 1.8, constant E_2 is 1.33, constant χ_q is 0.2, and

κ is the von Karman constant. The last term in Equation (9.6) accounts for the effects of solid walls and the free surfaces on the length scale (Mellor and Yamada 1982). The vertical eddy viscosity (ε_z) is defined as $\varepsilon_z = ql\chi_z$. The coefficients χ_z are the stability functions related to the Richardson number and expressed as:

$$\chi_z = \frac{A_2\left(1 - 6A_1/B_1\right)}{1 - 3A_2G_H\left(6A_1 + B_2\right)} \tag{9.7}$$

where $G_H = 0$ without considering the variation of flow density and the constants used in Equation (9.7) are $A_1 = 0.92$, $A_2 = 0.74$, $B_1 = 16.6$, and $B_2 = 10.1$ (Mellor and Yamada 1982).

9.1.1.2 Conceptual model

The conceptual model of a levee embankment strengthened by HPTRM on the crest and along the landward-side slope had the same geometry as the levee embankment built during the full-scale overtopping tests explained in Chapter 4. The dimensions of the levee embankment in the conceptual model were 26.12 m long × 3.25 m high × 3.66 m wide. The distance between the toe of the sea-side slope and the upstream boundary was 39.8 m. The width of the levee crest along the flow direction was 2.57 m. The sea-side has a slope of 1V:4.25H, and the landward-side slope has a slope of 1V:3H. To reduce the impact of the reflection of waves at the downstream boundary, a sufficiently large reservoir was used at the downstream boundary.

9.1.1.3 Boundary conditions

Stresses and vertical velocity were assigned for the boundary conditions at the free water surface and solid boundary on the levee embankment, which can be expressed by:

$$\rho\varepsilon_z\left(\frac{\partial u}{\partial x}, \frac{\partial v}{\partial x}\right) = \begin{cases} (0,0) & \text{at water surface} \\ \left(\tau_{bx}, \tau_{by}\right) & \text{on the levee} \end{cases} \tag{9.8}$$

$$w = \begin{cases} \dfrac{\partial \zeta}{\partial t} + u\dfrac{\partial \zeta}{\partial x} + v\dfrac{\partial \zeta}{\partial y} & \text{at water surface} \\ -u\dfrac{\partial H'}{\partial x} - v\dfrac{\partial H'}{\partial y} & \text{on the levee} \end{cases} \tag{9.9}$$

The boundary conditions of turbulence energy and turbulence macroscale applied at the water surface and solid boundaries, respectively, are

$$q^2 = B_1^{2/3}u_*^2 \tag{9.10}$$

$$q^2 l = 0 \tag{9.11}$$

where u_* is the friction velocity associated with the bottom frictional stress (τ_{bx}, τ_{by}). The bottom frictional stresses can be determined by:

$$\tau_b = \rho C_D |U_b| U_b \tag{9.12}$$

where C_D is the drag coefficient and U_b is the velocity in the grid point closest to the solid boundary. Stephan and Gutknecht (2002) defined the drag coefficient as:

$$C_D = \left(\frac{1}{\kappa} \ln \frac{D - y''}{y''} + 8.5 \right)^{-2} \tag{9.13}$$

where D is the water depth, κ is the von Karman constant (=0.4), and y'' is the mean deflected plant height. For the flow passing over the grass, there are two flow zones. The first zone is within the grass and the second zone is above it. The flow zone above the grass is characterized by the logarithmic velocity profile and its zero is raised by the deflected plant height y'' (Figure 9.1). The deflected plant height (y'') is a critical parameter (length scale) in this model because it affects the drag coefficient C_D and the bottom frictional stresses. The plant covering increases the bed surface roughness and the vertical velocity distribution.

9.1.1.4 Initial conditions

The upstream water level was the same as the levee crest height and there was not any water along the landward-side slope and in the downstream reservoir. The wave was generated at 39.8 m away from the sea-side slope toe. The initial input conditions were 11 surge elevations (h_1 from +0.2 m to +1.2 m above the levee crest, $h_1 = -R_c$) in the storm surge overflow. For the combined wave overtopping and storm surge overflow, the initial input conditions were at three surge elevations ($h_1 = +0.3$, +0.6, and +0.9 m above the levee crest), 15 significant wave heights (H_{m0} from 0.15 to 1.8 m), and two peak wave periods ($T_p = 7$ and 10 s). This yielded a total of 30 unique conditions for combined wave overtopping and storm surge overflow.

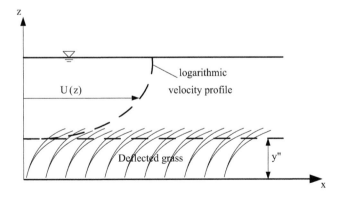

Figure 9.1 Vertical velocity profile above the grass. (Adapted from Li et al. (2015a). Reproduced with permission from Elsevier.)

9.1.1.5 Random wave generation

Random waves were applied as the upstream (sea-side) boundary condition for water level. The random waves were generated using the parameterized Joint North Sea Wave Project (JONSWAP) spectrum (Goda 1999) as:

$$S(f) = \frac{0.06238(1.094 - 0.01915\ln\gamma')}{0.230 + 0.0336\gamma' - 0.185(1.9 + \gamma')^{-1}} H_s^2 T_p^{-4} f^{-5} \exp\left(-1.25 T_p^{-4} f^{-4}\right) \gamma'^{\exp\left(-(T_p f - 1)^2/2\sigma^2\right)}$$

(9.14)

where $S(f)$ is the spectral density function, H_s is the significant wave height (= $H_{1/3}$; defined as an average of highest 1/3 waves), T_p is the peak wave period, f is the wave frequency, γ' is the spectral enhancement factor, and σ is 0.07 for $T_p f \leq 1$ or 0.09 for $T_p f > 1$. The spectral enhancement parameter γ' ranges 1–6 and has a normal distribution with a mean of 3.3 and a standard deviation of 0.79 (Hasselmann 1973).

In wave overtopping studies, H_s is usually replaced with energy-based significant wave height (H_{m0}) (Hughes and Nadal 2009). H_{m0} is also employed as the representative wave parameter and calculated by analyzing the generated random wave from Equation (9.14) in this study. The H_{m0} is defined as:

$$H_{m0} = 4.004\sqrt{m_0}$$

(9.15)

where m_0 is the 0-th spectral moment. The m_0 can be solved by:

$$m_0 = \int_0^\infty E(f)df$$

(9.16)

where $E(f)$ is the spectral energy density and f is the frequency.

9.1.1.6 Numerical scheme

The σ-coordinate transformation, which was developed by Lu and Wai (1998), was used to solve the hydrodynamic equations and mass conservation and transport equations in three substeps. The first step is solving the advection term in momentum equations (9.2) and (9.3) and kinetic energy of turbulence and macroscale of turbulent eddy by the Eulerian-Lagrangian method (Equations 9.5 and 9.6). The second step is approximating the horizontal diffusion terms in the momentum equations is via the implicit finite element method. Then, the last step is solving the vertical diffusion terms in the momentum equations, the vertical dispersion terms, and source-sink terms in the scalar transport equations by the implicit finite difference method.

9.1.2 Sensitivity analyses

The size of the domain and the level of discretization were determined based on a sensitivity analysis. This analysis identified the domain size needed so that overtopping hydraulics was not affected by the boundaries and the grid spacing. The grid spacing for lateral flow (y) direction was 0.25 m with a total of 16 cells. Because the water thickness changes along the flow direction on the crest and landside slope, the grid spacing for vertical flow (z) direction varied from 0.008 (crest) to 0.025 m (landslide slope) with a total of 10 cells. In the longitudinal (flow) direction, the cell size varied from 0.05 to 0.25 m. The three grid setups were almost the same in the flow (x) direction except for the different grid resolution on the crest and landward-side slope (Figure 9.2). The coarse grid was 0.25 m long (x-direction) with a total of 10 cells on the crest and 38 cells on the landward-side slope. The medium cell was 0.1 m long with a total of 25 cells on the crest and 96 cells on the landward-side slope. The fine grid was 0.05 m long with a total of 50 cells on the crest and 192 cells on the landward-side slope. One test was performed using the three-dimensional (3D) hydrodynamic model (Equations 9.1–9.6) to compare the simulation results using the three grid systems for the storm surge overflow with an inflow surge depth of 0.3 m.

Figure 9.3 shows the comparison of predicted overtopping discharge and streamwise velocity from the beginning of the crest to the toe of the landward-side slope for the three grid systems. The overtopping discharge and the streamwise velocity were sensitive to the grid resolution, specifically near the transition from the crest to the landward-side slope ($x = 55.9$ m). The coarse grid did not simulate the complex flow structure near the transition very well, whereas the fine and medium grid provided similar results. Hence, the medium grid was considered optimal and was selected for the simulations.

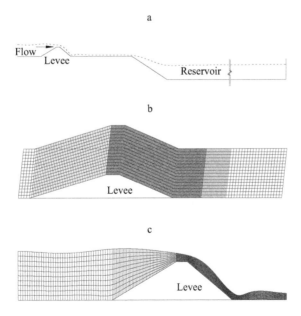

Figure 9.2 (a) Profile of levee embankment and downstream reservoir, (b) oblique view of the amplified horizontal orthogonal grid, (c) amplified vertical sigma grid. (Adapted from Li et al. (2015). Reproduced with permission from Elsevier.)

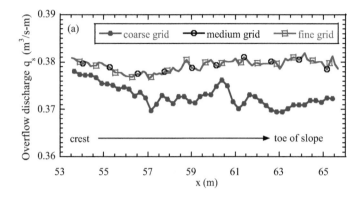

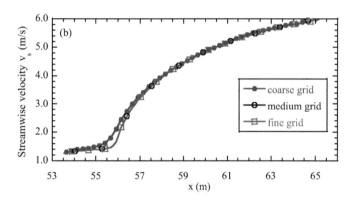

Figure 9.3 Comparison of predicted (a) steady overflow discharge and (b) streamwise velocity from the beginning of the crest to the toe of the landward-side slope in three grids (zero value of *x* was placed at the wavemaker) for the storm surge overflow with an inflow surge depth of 0.3 m. (Adapted from Li et al. (2015). Reproduced with permission from Elsevier.)

9.1.3 Model calibration

The 3D model (POM-WAD) was first calibrated with experimental results for storm surge overflow. Then, the model was verified with some empirical formulas and available laboratory measurements for combined wave overtopping and storm surge overflow. The model was used to study the overtopping hydraulics of combined wave overtopping and storm surge overflow on the HPTRM-strengthened levee at the toe of landward-side slope in particular.

In this study, the experimental data from full-scale overtopping tests on the HPTRM-strengthened levee were used to calibrate the unknown hydrodynamic parameters, the deflected plant height (y''), and to verify the validation of the 3D model.

Grass covering and rough mat increase the levee's bed surface roughness. The bed shear stress has a dominant effect on the momentum balances that control the magnitude and vertical structure of wave-driven flow and its distribution. As shown in

Figure 9.1, the deflected plant height (y'') is a key parameter to determine the drag coefficient, bed shear stress, and vertical velocity profile.

The model calibration was performed by adjusting the deflected plant height to match predicted flow velocity, flow thickness, and overflow discharge to the experimental measurements. The surge-only overflow experimental results with a freeboard ($-R_c$) of 0.3 m were used in the model calibration. Based on the experimental observation, the height of deflected grass was almost unchanged during the entire overtopping testing. Because the flow velocity was very fast, it was difficult for deflected grass to recover. The average value of the measured deflected plant height (y'') was 1.25 cm. The measured flow velocity, flow thickness, and overflow discharge at the two survey locations on the levee crest and the landward-side slope are listed in Table 9.1. The modeling predictions of flow velocity, flow thickness, and overflow discharge for the storm surge overflow with a surge height of 0.3 m are also shown in Table 9.1. Results showed that simulated flow velocity (V) and flow thickness (D) agreed well with the experimental observation. This suggests that the deflected plant height of 1.25 cm is a reasonable value for the mean deflected height of the grass in the hydrodynamic simulation of levee overtopping.

9.1.4 Storm surge overflow discharge

A total of 11 cases of storm surge overflow were simulated with the surge height of 0.2–1.2 m. Table 9.2 lists the input freeboard ($-R_c$) and the output discharges. The q_s is the steady overflow discharge per unit length.

Figure 9.4 shows the comparison of modeling predictions, laboratory data of storm surge overflow, and empirical solution (Henderson 1966) for the storm surge overflow discharge. The solid square points were the laboratory data measured in the full-scale storm surge overflow experiments. The modeling predictions were in general agreement with experimental measurements. The numerical modeling results and laboratory data were slightly lower than the broad-weir equation because of the increased resistance caused by the grass covering and rough mat.

Table 9.1 Comparison of model predictions and experimental measurements at the storm surge overflow with a surge height of 0.3 m

Location	Experiments of Chapter 6			Numerical model (present study)			Difference between the two methods		
	V (m/s)	D (m)	q_s (m³/s-m)	V (m/s)	D (m)	q_s (m³/s-m)	V (%)	D (%)	q_s (%)
Middle of crest	1.320	0.167	0.220	1.320	0.169	0.223	0	1.2	1.4
Landward-side slope	2.720	0.081	0.220	2.850	0.079	0.225	4.8	−2.5	2.3

Note: V is the mean flow velocity, D is the mean water thickness, q_s is the storm surge overflow discharge per levee width. The "middle of crest" is located at 0.91 m upstream of the interface between the crest and landward-side slope. The "landward-side slope" is located at 1.22 m downstream of the interface between the crest and landward-side slope.

Table 9.2 Hydrodynamic parameters and steady overflow discharge for storm surge overflows on the HPTRM-strengthened levee

Case number	Freeboard $-R_c$ (m)	Steady overflow discharge q_s (m³/s-m)
S1	0.2	0.102
S2	0.3	0.224
S3	0.4	0.373
S4	0.5	0.540
S5	0.6	0.723
S6	0.7	0.930
S7	0.8	1.151
S8	0.9	1.390
S9	1.0	1.643
S10	1.1	1.901
S11	1.2	2.184

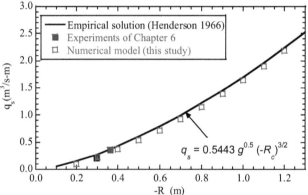

Figure 9.4 Estimation of steady overflow discharge using the modeling study, equation of Henderson (1966), and full-scale overtopping tests. (Adapted from Li et al. (2015). Reproduced with permission from Elsevier.)

9.1.5 Combined wave and storm surge overtopping discharge

The combined wave and storm surge overtopping flow is much more complex than storm surge overflow. Flow down the landward-side slope caused by combined waves and surge overtopping is unsteady and more difficult to analyze. In this study, a total of 30 simulations were run with freeboard ($-R_c$) ranging from 0.3 to 0.9 m, random waves with a JONSWAP spectrum, and energy-based significant wave height (H_{m0}) ranging from 0.2 to 1.8 m. Table 9.3 lists the input wave parameters for all 30 cases. Each simulation was performed to generate time sequences with a total duration of over 700 s (corresponding to approximately 100 waves).

Figure 9.5 shows an example of an instantaneous free surface output after 20, 22, 24, and 26 s. These free surfaces corresponded to the combined wave overtopping and storm surge overflow with the surge height of 0.6 m, energy-based significant wave height of 0.8 m, and peak wave period of 7 s.

Table 9.3 Hydrodynamic parameters for combined wave overtopping and storm surge overflow

Case number	Freeboard $-R_c$ (m)	Energy-based significant wave height H_{m0} (m)	Peak wave period T_p (s)	Relative freeboard $-R_c/H_{m0}$ (-)
SW1	0.3	0.381	7	0.787
SW2	0.3	0.528	7	0.569
SW3	0.3	0.566	7	0.530
SW4	0.3	0.641	7	0.468
SW5	0.3	0.778	7	0.386
SW6	0.3	0.897	7	0.334
SW7	0.3	1.014	7	0.296
SW8	0.3	1.143	7	0.262
SW9	0.3	1.277	7	0.235
SW10	0.3	1.570	7	0.191
SW11	0.3	2.312	7	0.130
SW12	0.3	3.132	7	0.096
SW13	0.3	3.840	7	0.078
SW14	0.3	0.778	7	0.386
SW15	0.3	1.570	10	0.191
SW16	0.3	0.778	10	0.771
SW17	0.6	1.014	7	0.592
SW18	0.6	1.570	7	0.382
SW19	0.6	2.030	7	0.296
SW20	0.6	2.312	7	0.259
SW21	0.6	3.132	7	0.192
SW22	0.6	3.840	7	0.156
SW23	0.6	4.660	7	0.129
SW24	0.6	1.143	7	0.787
SW25	0.9	1.570	7	0.573
SW26	0.9	2.312	7	0.389
SW27	0.9	3.132	7	0.287
SW28	0.9	3.840	7	0.234
SW29	0.9	4.660	7	0.193
SW30	0.9	0.381	7	0.787

Note: freeboard $(-R_c)$ is surge height above the crest.
From Li et al. (2015).

Figure 9.6 shows the cumulative total and mean overtopping volume calculated by the numerical model at the toe of the landward-side slope for the combined wave overtopping and storm surge overflow with the surge height of 0.3 m, energy-based significant wave height of 0.6 m, and peak wave period of 7 s. The units of overtopping volume were m^3/s per unit width of the meter. The unsteady nature of the overtopping events was clear (Figure 9.6). Figure 9.6 shows that the variation in the mean overtopping volume decreases rapidly and becomes very modest after the first 100 s.

The wave overtopping discharge rate is a critical parameter in the conceptual and preliminary design of levees. Based on the physical experiments and numerical models, several empirical formulas are available to predict the overtopping of levees under given wave conditions and storm surge levels (e.g., Reeve et al. 2008; Hughes and Nadal 2009; Rao et al. 2012a, 2012b). The overtopping discharge depends on wave parameters and structural parameters, including crest geometry, sea-side slope, landward-side slope, significant wave height, peak wave period, and water depth at the toe

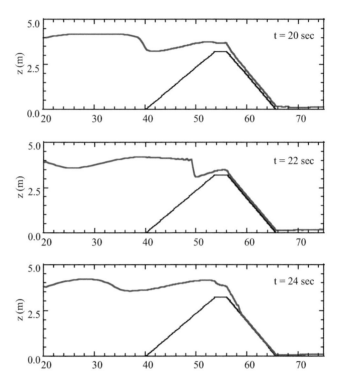

Figure 9.5 Instantaneous combined wave/surge overtopping process for one average period (Freeboard $-R_c = 0.6$ m, energy-based significant wave height
$H_{m0} = 0.8$ m, peak wave period $T_p = 7$ s). (Adapted from Li et al. (2015).
Reproduced with permission from Elsevier.)

of the landward-side slope. A detailed review of the overtopping dischargefor surge
overflow, wave overtopping, and combined wave and surge overtopping conditions
can be found in Hughes and Nadal (2009).

Figure 9.7 shows the dimensionless overtopping discharge for combined wave and
storm surge overtopping as a function of the relative (negative) freeboard. Figure 9.7
shows that the dimensionless overtopping discharge has a good trend with increasing
relative freeboard. According to Figure 9.7 the best-fit curve can be described as:

$$\frac{q_{ws}}{\sqrt{gH_{m0}^3}} = 0.034 + 0.457 \left(\frac{-R_c}{H_{m0}} \right)^{1.4} ; \quad R_c < 0 \tag{9.17}$$

The best-fit equation had a correlation coefficient of 0.996 and a root-mean-square
(RMS) error of 0.008. It should be noted that R_c must be entered as a negative number
so that the ratio in brackets would be positive. Like any empirical equation, application of Equation (9.17) should be limited to the assumed levee geometry that was a
sea-side slope of 1V:4.25H and a landward-side slope of 1V:3H.

Figure 9.7 shows the comparisons of overtopping discharges estimated from Equation (9.17), the predicted overtopping discharges on earthen levee using equations of

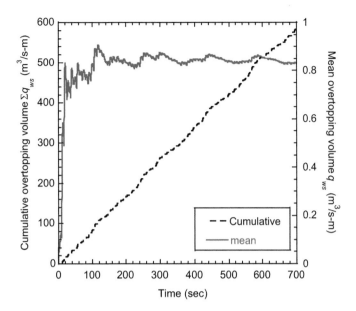

Figure 9.6 Time history of cumulative and mean overtopping volumes for combined wave overtopping and storm surge overflow at the toe of the landward-side slope. ($-R_c = 0.3$ m, $H_{m0} = 0.6$ m, and $T_p = 7$ s). (Adapted from Li et al. (2015). Reproduced with permission from Elsevier.)

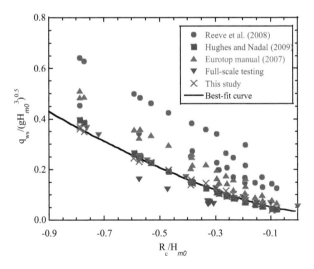

Figure 9.7 Comparisons of overtopping discharges estimated from the modeling study with the predicted overtopping discharges on earthen levee using equations of EurOtop manual (2007), Reeve et al. (2008), Hughes and Nadal (2009), and full-scale overtopping results explained in Chapter 6. (Adapted from Li et al. (2015). Reproduced with permission from Elsevier.)

EurOtop manual (2007), Reeve et al. (2008), Hughes and Nadal (2009), and the full-scale overtopping experimental results presented in Chapter 6. Figure 9.7 shows that the new equation can predict the overtopping discharge very well when compared with the full-scale overtopping experimental results and with Hughes and Nadal's equation (2009). Both equations of Revee et al. (2008) and EurOtop manual (2007) overpredicted the overtopping discharges, with equations of Revee et al. (2008) showing the greatest difference.

9.1.6 Flow parameters on the HPTRM-strengthened levee slope

The hydraulic conditions on the HPTRM-strengthened levee slope are complicated during the combined wave overtopping and storm surge overflow. There is no intuitive understanding of the distribution of flow parameters along the landward-side slope and at the toe of the landward-side slope. New empirical equations are available to characterize several representative parameters of wave-related unsteady flow on the landward-side levee slope (Hughes and Nadal 2009). The instantaneous flow thickness perpendicular to the levee slope (d) is a function of time (Figure 9.8). The peak flow thickness for each wave is designated as h, and the slope-perpendicular distance between the wave crest and the following wave trough is referred to as H. When the landward-side slope goes dry between overtopping waves, the h becomes H. For larger values of negative freeboard, the slope usually does not go dry with the passage of each wave. For these cases, wave trough flow thickness is positive and $h > H$ (Hughes and Nadal 2009). Figure 9.9 shows an example of the time series of flow thickness and flow velocity at the toe of the landward-side slope for the combined wave overtopping and storm surge overflow with the surge height of 0.3 m, the energy-based significant wave height of 0.78 m, and the peak wave period of 7 s. Unsteadiness and randomness of overtopping events were observed (Figure 9.9).

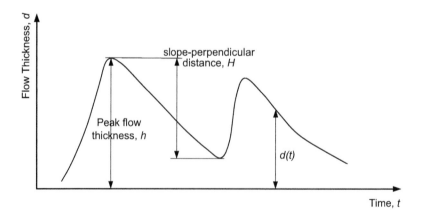

Figure 9.8 Definition sketch of flow thickness parameters on the landward-side slope. $d(t)$ is the instantaneous flow thickness perpendicular to the levee slope and h is the peak flow thickness for each wave, and H is the slope-perpendicular distance between the wave crest and the following wave trough. (Adapted from Li et al. (2015). Reproduced with permission from Elsevier.)

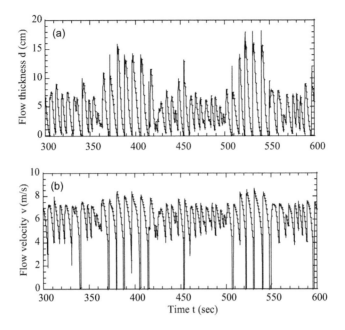

Figure 9.9 (a) Time history of flow thickness and (b) flow velocity at the toe of landward-side slope ($-R_c = 0.3$ m, $H_{m0} = 0.78$ m, and $T_p = 7$ s). (Adapted from Li et al. (2015). Reproduced with permission from Elsevier.)

9.1.7 Average flow thickness at the landward-side slope toe

The average flow thickness (d_m) was calculated by averaging the flow thickness at the end of the landward-side slope starting with data point 30 s and continuing till the end of the time series. An average was taken over the last four consecutive cells (4×0.1 m = 0.4 m) to represent the mean flow thickness at the toe of the landward-side slope. A correlation was sought between the mean flow thickness and the average overtopping discharge, and the results are shown in Figure 9.10. The best linear-fit equation given by simple empirical expressions are shown below:

$$\left(\frac{gR_c^3}{q_{ws}^2} \right)^{1/3} = 91.745 \left(\frac{d_m}{H_{m0}} + 0.145 \right)^2 - 1.941 \qquad (9.18)$$

Equation (9.18) has a correlation coefficient of 0.96 and a RMS error of 0.073. Hughes and Nadal (2009) provided a more detailed expression of average flow thickness along the landward-side slope as a function of the landward-side slope angle. Application of Equation (9.18) is limited to the landward-side slope of 1V:3H.

Figure 9.10 shows the predictions of the average water thickness at the toe given by Hughes and Nadal (2009). The numerical model results were approximately 37% smaller than the experimental results for average water thickness. This difference could be related to the different setup for the levees between their flume model and our numerical flume. There was a long and mild berm with the slope of 1:24 after the

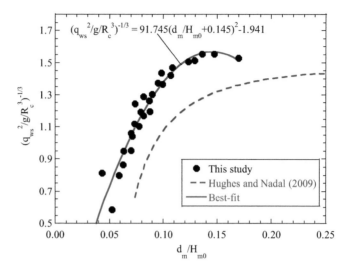

Figure 9.10 Mean flow thickness at the toe of the landward-side slope as a function of average overtopping discharge. (Adapted from Li et al. (2015). Reproduced with permission from Elsevier.)

landward-side slope in the Hughes and Nadal's (2009) flume, while this was not the case in the numerical experiment used in this study. The berm difference could cause the variation of water depth and velocity at the slope toe.

9.1.8 Time series upcrossing analysis

The calculated time series of flow thickness at the toe of the landward-side slope were analyzed by the zero-upcrossing technique to identify the maximum and minimum flow thicknesses for all waves contained between data points 30 and 700 s. The characteristic slope-perpendicular wave height parameters H_{rms} and $H_{1/3}$ were determined for each measured time series.

H_{rms} represents the RMS slope-perpendicular wave height calculated using the statistic maximum and minimum flow thicknesses, as shown in Equation (9.19) below:

$$H_{rms} = \sqrt{\frac{1}{n}\sum_{m=1}^{n} H_m^2} \qquad (9.19)$$

where n is the amount of random wave, and H_m is the slope-perpendicular wave height of the m_{th} wave. A relationship was sought to determine a relationship for H_{rms} in terms of other parameters that could be specified. Figure 9.11 shows the normalized average flow thickness for various values of the relative freeboard. The solid curve represents the best-fit of a linear function described as follows:

$$\frac{H_{rms}}{d_m} = 2.282 \cdot \exp\left(0.573 \cdot \frac{R_c}{H_{m0}}\right) \qquad (9.20)$$

This best-fit equation had a correlation coefficient of 0.927 and an RMS percent error of 0.056. The surge height (R_c) should be entered as a negative number and the equation should not be applied for cases where $R_c \geq 0$. Figure 9.11 compares the predictions made by Hughes and Nadal (2009) and this study's data. Results showed that the numerical results were smaller than that of experimental data due to the increased roughness effects of the grass cover.

Time-domain significant slope-perpendicular wave height parameter ($H_{1/3}$) was determined for each trial and calculated by the following equation:

$$H_{1/3} = \frac{3}{n} \sum_{m=1}^{n/3} H_m \tag{9.21}$$

A relationship was sought to express $H_{1/3}$ in terms of other parameters that could be specified. The significant slope-perpendicular wave height parameter is shown in Figure 9.12a for various values of RMS slope-perpendicular wave height. The solid curve represents the best-fit of a linear function given as:

$$H_{1/3} = 1.353 H_{rms} \tag{9.22}$$

This best-fit equation had a correlation coefficient of 0.999 and an RMS percent error of 0.013. The corresponding peak wave flow thickness (h_{rms}) and $h_{1/3}$ were also determined from the calculated time series via Equations (9.23) and (9.24) shown below:

$$h_{rms} = \sqrt{\frac{1}{n} \sum_{m=1}^{n} h_m^2} \tag{9.23}$$

$$h_{1/3} = \frac{3}{n} \sum_{m=1}^{n/3} h_m \tag{9.24}$$

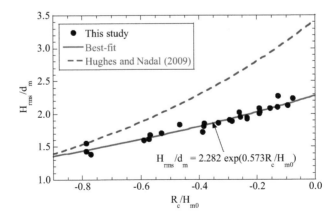

Figure 9.11 Prediction of H_{rms}/d_m on the landward-side slope as a function of relative freeboard R_c/H_{m0}. (Adapted from Li et al. (2015). Reproduced with permission from Elsevier.)

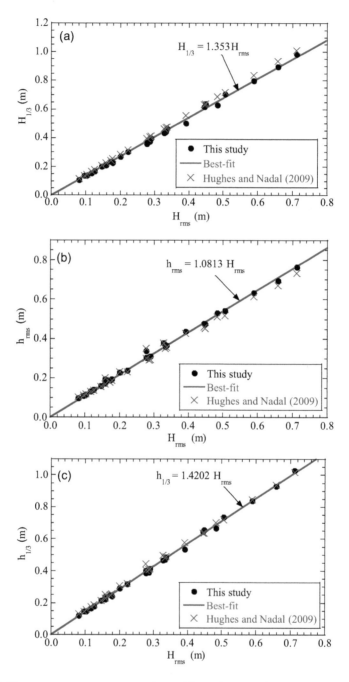

Figure 9.12 Time series upcrossing analysis: (a) prediction of landward-side slope toe significant slope-perpendicular wave height parameter $H_{1/3}$, (b) representative flow thickness H_{rms}, and (c) representative flow thickness $h_{1/3}$ as a function of the wave height parameter H_{rms}. (Adapted from Li et al. (2015). Reproduced with permission from Elsevier.)

where h_m is the peak wave flow thickness of the m_{th} wave. These values were obtained by considering only the wave peaks and not the troughs. Figure 9.12b and c show the two representative flow thickness parameters as a function of the wave height parameter H_{rms}.

The solid lines shown in the plots of Figure 9.12 are the best-fits to the data given by the following empirical equations:

$$h_{\text{rms}} = 1.0813 \cdot H_{\text{rms}} \tag{9.25}$$

$$h_{1/3} = 1.4202 \cdot H_{\text{rms}} \tag{9.26}$$

The correlation coefficients for Equations (9.25) and (9.26) were 0.999, and the corresponding RMS errors were 0.011 and 0.010, respectively. It should be noted that Equations (9.25) and (9.26) are only appropriate for combined wave overtopping and storm surge overflow.

Figure 9.12 shows that the numerical results of ratios between these wave height parameters are almost the same as the estimates by Hughes and Nadal (2009). Although the wave height was resisted by the long and rough slope, the shape of waves and the basic characteristics related to the ratio of wave heights at the slope toe did not change.

9.1.9 Estimation of wavefront velocity on the landward-side slope

Hughes and Nadal (2009) estimated the speed of the wavefront by using the time which took the wavefront to move from one pressure gauge to the next. The same method was applied in this study to estimate the wavefront velocity on the landward-side slope. Figure 9.13 shows a portion of the flow thickness time series at the beginning of the landward-side slope (solid curve) thathas been time-shifted based on the equivalent time series from the toe of the landward-side slope (dotted curve). In this case, the shift was 0.6 s over a down-slope distance of 4.5 m. Therefore, a rough estimate of the wavefront velocity was 7.5 m/s. This approximate wavefront velocity represents an average value along the landward-side slope. In fact, the flow may still be accelerating over this range. Nevertheless, the matching of shifted time series would seem to indicate that the waves had a nearly constant form as they propagate down the slope.

Figure 9.14 shows the estimated wavefront velocity as a function of the H_{rms}. The discrete jumps in the velocity were attributed to the subjective judgment of the superposition of two time series. A similar phenomenon was also observed in Hughes and Nadal (2009). The straight line shown in Figure 9.14 is the best-fit given by the following simple equation:

$$v_m = 3.05 + 3.32 \left(g H_{\text{rms}} \right)^{1/2}, \text{ for } R_c \geq 0.3 \, \text{m and } H_{m0} \geq 0.381 \, \text{m} \tag{9.27}$$

where v_w is the wavefront velocity. This best-fit had a correlation coefficient of 0.968 and an RMS percent error of 0.312. The coefficient in Equation (9.27) is constant for this particular data set, but it may be a function of slope angle and surface roughness.

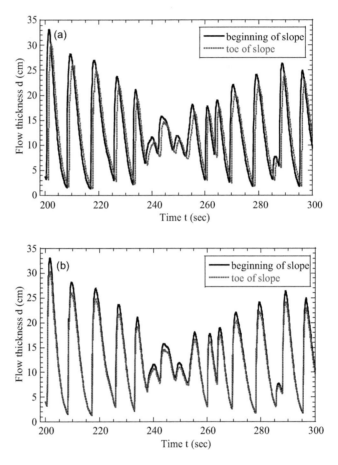

Figure 9.13 Flow thickness time series at the beginning and the toe of the landward-side slope: (a) before time-shifted and (b) after time-shifted (data from trial 17). (Adapted from Li et al. (2015). Reproduced with permission from Elsevier.)

9.2 Smoothed Particle Hydrodynamic (SPH) method

SPH is a meshless fully Lagrangian method for obtaining numerical solutions for the equations of fluid dynamics by replacing the fluid with a set of particles (e.g., Monaghan 2005). In this approach, the particles are interpolation points from which fluid properties can be calculated. The SPH particles are also material particles which can be treated like any other particle system. The advantages of SPH include (1) no meshor grid is required, and SPH can deal with large deformations of the free surface and interface problems; (2) no mesh refinement for any change in density, viscosity, and flow morphology is needed; (3) no gap between the continuum and fragmentation for brittle fracture and subsequent fragmentation in damaged solids exists; and (4) computational advantage (e.g.; Monaghan 2006; Li et al. 2012; Rao et al. 2012;

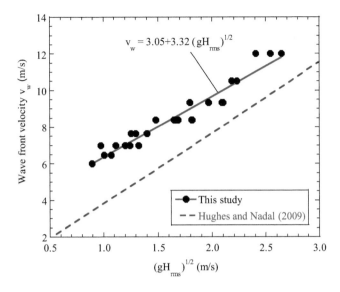

Figure 9.14 Estimated wavefront velocity on the landward-side slope. (Adapted from Li et al. (2015). Reproduced with permission from Elsevier.)

Li et al. 2013, 2015a). The computational advantage feature of the SPH is very useful for two-phase flow with water–structure interactions.

9.2.1 Numerical methodology

9.2.1.1 SPH method

To find the value of a particular quantity f at an arbitrary point (x), interpolation is applied as follows (Monaghan 2005):

$$f(x) = \sum_j f_j w(x - x_j) V_J \qquad (9.28)$$

where f_j is the value of f associated with particle j located at x_j, $w(x - x_j)$ is the weighting of the contribution of particle j to the value of $f(x)$ at position x, and V_J is the volume of particle j which is defined as the mass (M_J) divided by the density of the particle (ρ_j).

The Kernel $w(x - x_j)$ is a smoothing function and varies with the distance from x. When the Kernel smoothing length (h) and interparticle spacing (Δx) are smaller, the Kernel is assumed to have compact support so that the sum is only taken from neighboring particles.

The SPH method usually considers the fluid as compressible and directly calculates the pressure from an equation of the state. The conservation of mass and the conservation of the moment are written in particle form as follows (Monaghan 1992):

$$\frac{du_i}{dt} = -\sum_j m_j \left(\frac{P_j}{\rho_j^2} + \frac{P_i}{\rho_i^2} \right) \cdot \nabla w(x_i - x_j) \qquad (9.29)$$

where u_i is the velocity of the particle, ρ_i is the density of the particle, P_j is the pressure at the particle, and m_j is the mass of the particle (j). Π_{ij} is an empirical term representing the effects of viscosity and calculated as follows (Monaghan 1992):

$$\Pi_{ij} = \begin{cases} -\alpha_{ij}\mu_{ij} + \beta\tilde{c}_{ij}\mu_{ij}^2 & (u_i - u_j)(x_i - x_j) < 0 \\ 0 & \text{elsewhere} \end{cases} \tag{9.30}$$

where α and β are empirical coefficients, $\tilde{c}_{ij} = (c_i + c_j)/2$ and $\mu_{ij} = \ell(u_i - u_j)(x_i - x_j)/(\gamma_{ij}^2 + 0.01h^2)$. The parameters α and β are often assumed as 0.01 and -0.1, respectively. Thus, the stabilized discrete momentum equations become:

$$\frac{du_i}{dt} = -\sum_j m_j \left(\frac{P_j}{\rho_j^2} + \frac{P_i}{\rho_i^2} + \Pi_{ij} \right) \nabla w(x_i - x_j) \tag{9.31}$$

The fluid was treated as slightly compressible in this method, following Monaghan and Kos (1999), Colagrossi and Landrini (2003), and Dalrymple and Rogers (2006). Based on Batchelor (1974), the relationship between pressure and density was assumed as:

$$p = B\left[\left(\frac{\rho}{\rho_0} \right)^\gamma - 1 \right] \tag{9.32}$$

where $\gamma = 7$ and $B = 100\rho_0 H/\gamma$, with water density $(\rho_0) = 1000\,\text{kg/m}^3$, and the water depth ($H$).

9.2.1.2 Conceptual model

The conceptual model of a levee embankment strengthened by RCC on the crest and along the landward-side slope is shown in Figure 9.15. The width of the levee crest along the flow direction was 2 m. The seaward-side had a slope of 1V:4.25H, and the landward-side had a slope of 1V:3H (Rao et al. 2012).

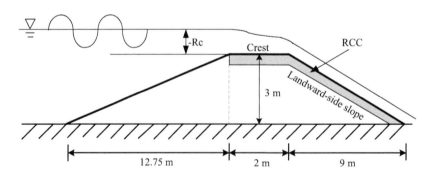

Figure 9.15 Conceptual setup of earthen levee-strengthened by RCC on the crest and along the landward-side slope. (Adapted from Rao et al. (2012). Reproduced with permission from Elsevier.)

9.2.1.3 Boundary conditions

One of the major boundary treatment methods in SPH is the ghost particle (type I) technique (Liu et al. 2002). The principle utilized when dealing with the boundaries is to make the Kernel have a compact support (i.e., the smoothing circle is full of particles) (Figure 9.16). Near the boundary, the Kernel does not have a complete smoothing circle. As the water boundary is moving forward, symmetrical ghost particles are added so that the smoothing circle is full. Depending on the boundary conditions, ghost particles were assigned different velocity, mass, and volume properties.

If the boundary is solid, another type of virtual particles (type II) is established on the boundary. When the real particles approach the solid boundary, a force is produced to avoid interpenetration of particles. This force acts along the centerline between the virtual particles and the real particles and points to the real particles (Figure 9.17).

In this chapter, these two types of boundary treatments were used. Ghost particles (type I) were assigned along the entire boundary. Negative water depth and velocity of real particles were used for the ghost particles to keep the pressure and velocity of the boundary equal to zero. Solid boundary particles (type II) were assigned along the ground surface and the levee profile. Because the RCC is located on the crest and landward-side slope of the levee, the solid boundary particles on this part required special treatment, and each particle in this area had its own mass, position, and volume. These solid particles provide a friction force that flows at the same time as the resistance

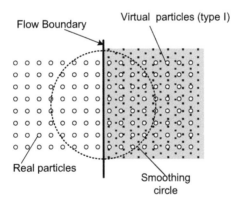

Figure 9.16 Illustration of the ghost particles method. (Adapted from Rao et al. (2012). Reproduced with permission from Elsevier.)

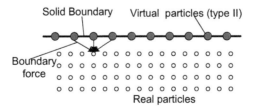

Figure 9.17 Illustration of type II virtual particles on a solid boundary. (Adapted from Rao et al. (2012). Reproduced with permission from Elsevier.)

force. The friction force was determined by the RCC material and was calculated by the following equation (Stephan and Gutknecht 2002):

$$\tau = \rho C_D \upsilon^2 \qquad (9.33)$$

where C_D is the equivalent roughness, ρ is the density of water, and v is the velocity of the flow particles. There was no available equivalent roughness for RCC. Therefore, the equivalent roughness of concrete of 0.0012 was taken (Gupta 2008).

9.2.1.4 Initial conditions

The initial particle setup is shown in Figure 9.18 (not including the type I ghost particles). There were 9,998 flow particles and 5,332 solid boundary particles (type II) in the computational domain. The change in the number of ghost particles (type I) depends on time. The surge overflow depth in the upstream was assumed to be constant during the computational time after achieving the steady-state surge overflow conditions. The wave was generated at 20 m away from the seaward-side slope toe. The initial input conditions were three surge elevations ($h_1 = +0.2$, $+0.4$ and $+0.6$ m above the levee crest), three significant wave heights ($H_{m0} = 0.3$, 0.6 and 0.9 m), and three peak wave periods ($T_p = 4$, 5, and 6 s). This yielded a total of 27 unique conditions for combined wave and surge overtopping.

Six positions on the section were chosen to collect the hydraulic data (Figure 9.18). The particles across sampling stations S1–S6 were used to calculate the flow thickness, horizontal velocity, and overtopping discharge.

9.2.2 Numerical wave generator

Waves were generated by specifying a velocity profile for the particles located in the wave generator part. The waves were generated by the following equations (Rune et al. 2011):

$$\text{Horizontal velocity}: \ u = \frac{a \cdot n \cdot \cos hk(z+H)}{\sin hkH}\cos(kx - nt) + u_{overtopping} \qquad (9.34)$$

$$\text{Vertical velocity}: \ w = \frac{a \cdot n \cdot \sin hk(z+H)}{\sin hkH}\sin(kx - nt) \qquad (9.35)$$

where a is the velocity scaling parameter, n is the circular rate frequency scaling parameter, $u_{overtopping}$ is used to maintain the specified still water elevation for overtopping, z is the water elevation, H is the water height, x is the horizontal location, k is the wavenumber, and t is time.

9.2.3 Sensitivity analysis

Sensitivity analyses were conducted to assess how the SPH model is affected by two key parameters, including particle distance (Δx) and smoothing length (l). The particle distance (Δx) and smoothing length (l) were varied for one parameter at a time while keeping all other parameters the same as the base case. Predictions were made

(a)

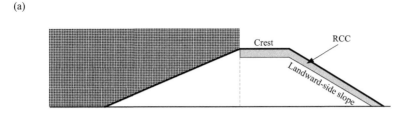

(b)

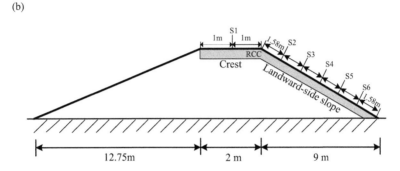

Figure 9.18 (a) Initial particles setup and (b) position of data collection. (Adapted from Rao et al. (2012). Reproduced with permission from Elsevier.)

for the surge-only overflow conditions with an inflow surge depth of 0.4 m. The steady discharge (q_s) was compared to the analytical solution from the open-channel flow equation (Henderson 1966) as:

$$q_s = 0.5443\sqrt{g}h_1^{3/2} \tag{9.36}$$

where q_s is the steady discharge, h_1 is the surge depth.

Table 9.4 shows that the SPH model is sensitive to the particle distance (Δx). When the Δx was set as 0.05 m, the predicated steady discharge was closed to the analytical prediction made by the Equation (9.36). However, the simulation time was nearly four times longer than that of the base case. For accuracy and reasonable simulation time, the particle distance Δx was used as 0.1 m in this study. For the variation of smoothing length (l), the SPH predicted discharges were almost constant when the length was longer than $2\Delta x$. These results suggested using smoothing length as twice as the particle distance ($l=2\Delta x$) in this study.

9.2.4 Average overtopping discharge for combined wave and surge overtopping

Table 9.5 lists the input wave parameters of all 27 cases and the output discharges (q_s and q_{ws}). The q_s is the surge-only steady overflow discharge per unit length, and the q_{ws} are combined wave and surge overtopping discharge per unit length. The q_s is the

Table 9.4 Results of sensitivity analyses for the SPH model

Case number	Input variables		Result		
	Particle distance Δx (m)	Smoothing length l	SPH predicted q_s (m^3/s-m)	Analytical solution of q_s (m^3/s-m)	CPU time (h)
1 (base case)	0.10	$2\Delta x$	0.40	0.44	14
2	0.20	$2\Delta x$	0.28	0.44	4
3	0.05	$2\Delta x$	0.42	0.44	45
4	0.10	Δx	0.38	0.44	12
5	0.10	$3\Delta x$	0.40	0.44	16

Table 9.5 Hydrodynamic parameters and average overtopping discharge

Case number	H_{m0} (m)	T_p (s)	q_{ws} (m^3/s-m)	q_s (m^3/s-m)
Surge level = +0.2 m above crest				
1	0.285	3	0.20	0.16
2	0.305	4	0.21	0.16
3	0.310	5	0.19	0.16
4	0.622	3	0.25	0.16
5	0.593	4	0.24	0.16
6	0.597	5	0.27	0.16
7	0.902	3	0.34	0.16
8	0.912	4	0.33	0.16
9	0.880	5	0.36	0.16
Surge level = +0.4 m above crest				
10	0.285	3	0.41	0.40
11	0.305	4	0.43	0.40
12	0.310	5	0.36	0.40
13	0.622	3	0.46	0.40
14	0.593	4	0.49	0.40
15	0.597	5	0.45	0.40
16	0.902	3	0.52	0.40
17	0.912	4	0.50	0.40
18	0.880	5	0.51	0.40
Surge level = +0.6 m above crest				
19	0.285	3	0.81	0.80
20	0.305	4	0.77	0.80
21	0.310	5	0.83	0.80
22	0.622	3	0.84	0.80
23	0.593	4	0.79	0.80
24	0.597	5	0.82	0.80
25	0.902	3	0.84	0.80
26	0.912	4	0.87	0.80
27	0.880	5	0.83	0.80

Note: H_{m0} is the energy-based significant wave height, and T_p is the peak spectral wave period. q_s is the surge-only steady overflow discharge per unit length, and q_{ws} is the combined wave and surge overtopping discharge per unit length.

After Rao et al. (2012).

average of the first 50 discharge data points taken as the steady overflow discharge at S1, and q_{ws} is the average for 500 instantaneous discharge data points at S1. These 500 data points did not include any data from the initial steady overflow portion before waves arriving at the levee.

Figure 9.19 provides an example of output for instantaneous flow thickness, horizontal velocity, and discharge for a 10–22 s time series at location S1. The wave overtopping discharge rate is a critical parameter in the conceptual and preliminary design of levees. Based on physical experiments and numerical models, several empirical formulas have been provided by researchers to predict the overtopping of levees under given wave conditions and water levels (Reeve et al. 2008; Hughes and Nadal 2009). The overtopping discharge depends on wave parameters and structural parameters, including the seawall freeboard, crest geometry, seaward slope, significant wave height, mean or peak wave period, angle of wave attack measured from the normal to the structure, water depth at the toe of the seawall, and seabed slope. Hughes and Nadal (2009) provided a detailed review of the overtopping discharge for surge overflow, wave overtopping, and combined wave and surge overtopping conditions.

Figure 9.20 shows the dimensionless combined wave/surge average overtopping discharge versus the relative freeboard for all 27 cases. The best-fit curve yields a good trend with increasing relative freeboard. The solid line is a best-fit empirical equation given by the formula:

$$q_{ws} / \sqrt{gH_{m0}^3} = 0.04 + 0.57(-R_C/H_{m0})^{1.35} \qquad (9.37)$$

where q_{ws} is the combined wave and surge overtopping discharge per unit length, H_{m0} is the energy-based significant wave height, g is the gravitational acceleration constant, and freeboard R_c is defined as the vertical distance between the still water elevation and crest elevation.

It should be noted that R_c must be entered as a negative number so that the ratio in brackets would be positive. It was determined that the peak spectral wave period had a negligible influence on the determination of q_{ws} for the range of periods tested in the model. This finding is consistent with Hughes and Nadal (2009). The application of Equation (9.37) is limited to the assumed levee geometry, which is a seaward-side slope of 1:4.25 (V:H), and the roughness of the protective system being assumed similar to RCC.

During a storm surge when the still water level is higher than the levee crest (i.e., negative freeboard with $R_c < 0$), water overflow of the levee occurs and subcritical flow exists on the high water-side of the levee. Critical flow (transition between subcritical and supercritical flow) generally occurs on the crest near the landward-side edge and the flow down the landward-side slope is supercritical unless the slope is very flat. As the still water level remains unchanged, the overflow discharge will remain steady. When frictional energy loss along the crest is negligible, the discharge per unit width of levee can be computed by the generally accepted equation for flow over a broad-crested weir given by Equation (9.36).

Equation (9.37) can be written in terms of the steady overflow discharge by recognizing that $h_1 = -R_c$ and substituting Equation (9.36) to yield the following equation:

$$q_{ws} = 0.04\sqrt{gH_{m0}^3} + 0.95q_s\left(-R_C/H_{m0}\right)^{0.07}; \ R_C < 0 \qquad (9.38)$$

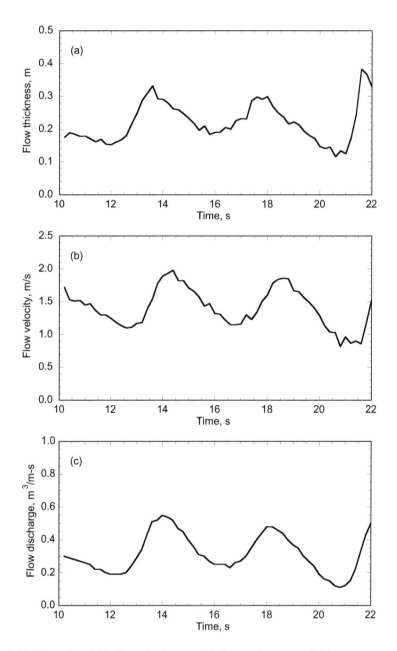

Figure 9.19 Example of (a) flow thickness, (b) flow velocity, and (c) overtopping discharge at the location of SI. (Adapted from Rao et al. (2012). Reproduced with permission from Elsevier.)

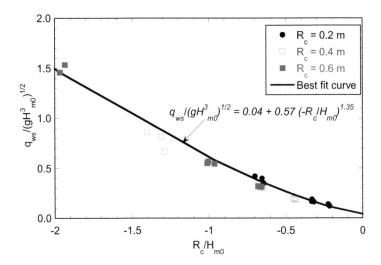

For higher absolute values of relative freeboard, the second term of Equation (9.38) dominates and the average combined wave/surge overtopping becomes approximately equal to the steady overflow discharge (q_s). Figure 9.21 shows that the ratio of q_{ws}/q_s is slightly greater than unity except for low values of negative freeboard, where the wave overtopping is more influential. Similar results were also reported by Hughes and Nadal (2009).

Figure 9.22 presents the estimates from the empirical equations of Reeve et al. (2008), Schüttrumpf et al. (2001), Hughes and Nadal (2009), and full-scale experimental data (Chapter 6). Hughes and Nadal (2009) developed a simple relationship between the combined wave and surge discharge and relative freeboard based on 27 unique wave conditions in a 25:1 scale laboratory study. Schüttrumpf et al. (2001) added the wave-only overtopping and surge-only overflow discharge to estimate the combined wave and surge overtopping discharge based on their laboratory study. Reeve et al. (2008) developed equations for the dimensionless average discharge of combined wave and surge overtopping based on numerical modeling.

Compared to the full-scale measurement and Hughes and Nadal's (2009) model, the SPH method overestimated the relative discharge for the combined wave and surge overtopping when the relative freeboard is less than −0.2. This overestimation could be related to the equivalent roughness of concrete used in Equation (9.33), which was not measured equivalent roughness for RCC. The SPH model estimated discharge by counting the number of particles flowing through a certain location. Because some particles could not fully pass through the location, the discharge could be overestimated. The predictions made by Schüttrumpf (2001) and Reeve et al. (2008) were too far from the measured data, which was also observed by Hughes and Nadal (2009).

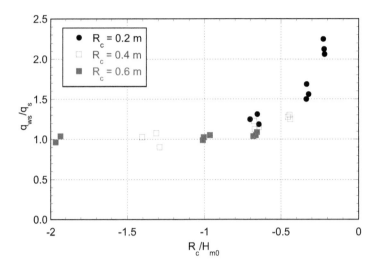

Figure 9.21 Influence of wave height on the discharge of combined wave and surge overtopping for RCC tests. (Adapted from Rao et al. (2012). Reproduced with permission from Elsevier.)

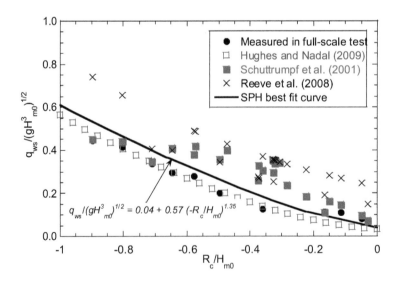

Figure 9.22 Estimation of overtopping discharge using full-scale overtopping tests, SPH best-fit equation, and equations of Schüttrumpf et al. (2001), Reeve et al. (2008) and Hughes and Nadal (2009). (Adapted from Rao et al. (2012). Reproduced with permission from Elsevier.)

9.2.5 Flow parameters on the landward levee slope

Flow down the landward-side slope caused by combined wave and surge overtopping is an unsteady and more complex process. Thus, it requires additional parameters for a thorough analysis. In this chapter, empirical equations were developed to characterize three representative parameters for wave-related unsteady flow on the landward-side slope. These parameters are defined on a time history diagram of an overtopping wave (Figure 9.8). The instantaneous flow thickness perpendicular to the levee slope (d) is a function of time. The peak flow thickness for each wave is designated as h, and the slope-perpendicular distance between the wave crest and following wave trough is referred to as H. If the landward-side slope goes dry between overtopping waves, then h becomes H. For larger values of negative freeboard, the slope usually does not go dry with the passage of each wave. For these cases, wave trough flow thickness is positive and $h > H$.

The output data on the landward-side levee slope were time series of flow thickness (d_m) at the five locations of S2–S6 (Figure 9.18b). Figure 9.23 shows the changes in flow thickness with a short time series for S2–S6 on the landward-side slope from one combined wave and surge overtopping test. It should be noted that the time difference between the locations for the wave leading edge is nearly uniform. This indicates that the wavefront is moving at a nearly constant speed down the slope, and this aspect is examined in more detail in the following section.

9.2.5.1 Average flow thickness on landward-side slope

The average flow thickness (d_m) was calculated at each location starting with data point 100 and continuing to the end of the time series. The RMS elevation was also determined. An average was taken over S3–S6 to represent the mean flow thickness on the slope. A correlation was sought between d_m and the hydrodynamic force and a reasonable result was obtained (Figure 9.24).

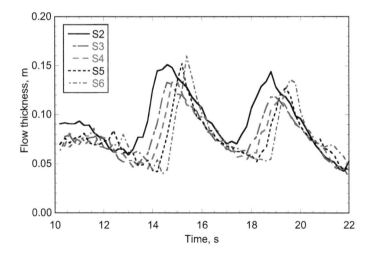

Figure 9.23 Example of flow thickness time series from S2 to S6. (Adapted from Rao et al. (2012). Reproduced with permission from Elsevier.)

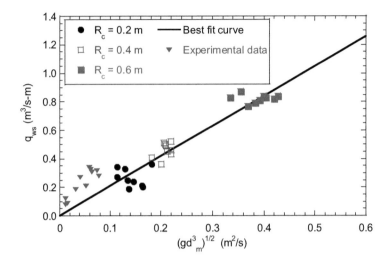

Figure 9.24 Average overtopping discharge versus mean flow thickness along the landward-side slope. (Adapted from Rao et al. (2012). Reproduced with permission from Elsevier.)

The solid line is a linear best-fit equation and shown below:

$$q_{ws} / \sqrt{gd_m^3} = 2.4 \tag{9.39}$$

If the mean velocity on the landward-side slope is defined as $v_m = q_{ws}/d_m$, then

$$v_m = 2.4\sqrt{gd_m} \tag{9.40}$$

which bears a resemblance to the Chezy equation.

The Chezy equation for wide channels (hydraulic radius approximately equals flow thickness, (d)), steep slopes, and steady flow (friction slope equals bed slope) is given as:

$$v = C_Z \sqrt{d \sin\theta} \tag{9.41}$$

where C_z is the Chezy coefficient and θ is the landward-side slope angle. The Chezy coefficient can be expressed in terms of the Fanning friction factor (f_F) as:

$$C_Z = \sqrt{\frac{2g}{f_F}} \tag{9.42}$$

Substituting Equation (9.42) into Equation (9.41) yields

$$v = \sqrt{\frac{2\sin\theta}{f_F}} \cdot \sqrt{gd} \tag{9.43}$$

Based on the assumption that the Chezy equation is an appropriate model to approximate the average of the rapidly varying unsteady flow situation modeled, it is hypothesized that the constant in Equations (9.40) and (9.41) is a function of both levee slope and a representative friction factor, as given by the first radical term in Equation (9.43). The Chezy coefficient varies with flow thickness on the slope; therefore, the friction factor (f_F) also varies continuously during the overtopping flow. By assuming the constant in Equation (9.41) is equal to the first radical term in Equation (9.43), the overall representative value for the friction factor for these experiments would be $f_F=0.0734$. For the 3V:1H landward-side slope, the constant in Equations (9.40) and (9.41) was assumed to be $2.4=4.61\sqrt{\sin\theta}$.

Substituting for the constant in Equation (9.40) and rearranging gives a tentative equation for the mean flow thickness as shown below:

$$d_m = 0.32\left[\frac{1}{g\sin\theta}\right]^{1/3}(q_{ws})^{2/3} \tag{9.44}$$

The mean flow velocity equation becomes the following:

$$v_m = 2.7\left(q_{ws}\cdot g\cdot\sin\theta\right)^{1/3} \tag{9.45}$$

The constants in Equations (9.44) and (9.45) are related to the slope roughness, and the equations are strictly applicable only for the landward-side slopes of 3V:1H having a roughness similar to the RCC system. Friction factors for naked levee slopes may not be much higher than RCC, but slopes armored with riprap or similar material will likely have significantly higher representative friction factors. Schüttrumpf and Oumeraci (2005) noted that the friction factor is an influential parameter for wave-only flows over levees and dikes and more research is needed to determine appropriate representative friction factors for a range of slope roughness.

9.2.5.2 Estimation of H_{rms} on the landward-side slope

H_{rms} is the RMS slope-perpendicular wave height. Values of H_{rms} at S3–S6 on the landward-side slope indicated little variation. A relationship was sought to express H_{rms} in terms of other parameters that could be specified. Figure 9.25 presents the best correlation relationships. Wave period was determined to have only a marginal influence. The solid curve represents the best-fit of a one-parameter exponential function shown below:

$$H_{rms}/d_m = 2.8\exp\left(-R_C/H_{MO}\right); \ R_C < 0 \tag{9.46}$$

It should be noted that when applying Equation (9.46), R_c must be entered as a negative number and the equation should not be applied for cases when $R_c \geq 0$ and the landward-side slopes are different from 3V:1H.

9.2.5.3 Estimation of wavefront velocity on the landward-side slope

Flow velocity was calculated using the time it took the wavefront to move from one location point to the next. The accuracy of wavefront velocity estimates using this

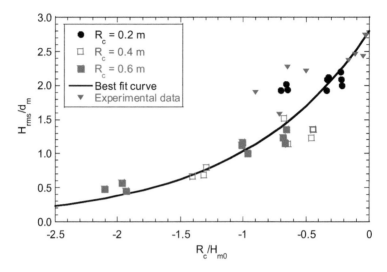

Figure 9.25 Estimation of H_{rms} on the landward-side slope as a function of d_m. (Adapted from Rao et al. (2012). Reproduced with permission from Elsevier.)

technique is limited by the data sampling. For each case, the entire time series between data points 1,000 and 15,000 at S3 and S6 were uniformly time-shifted one step at a time until the sum of the squares of differences between the two signals was minimized. The time shift associated with the minimum difference was used to estimate the wavefront velocity. As shown in Figure 9.23, the time shift over a down-slope distance of 6 m from S2 to S6 was 1.1 s. This approximate wavefront velocity represented an average value over the distance between S2 and S6.

Figure 9.26 plots the estimated wavefront velocity versus $\sqrt{gH_{rms}}$. Values of H_{rms} were taken from S3 and S6. The full-scale experiment data points are also shown. Although the experiment data and SPH results both showed a clear trend, the SPH wavefront velocity seemed to be higher than the actual measurements. This could be caused by the error in high-velocity flow measurement and the positioning of flow front particles in SPH. The straight line shown in Figure 9.26 is the best-fit line given by the simple equation:

$$v_m = 3.7\sqrt{gH_{rms}} \tag{9.47}$$

The coefficient in Equation (9.47) is constant for this particular data set, but it may be a function of slope angle and surface roughness, even though Equation (9.19) for estimating H_{rms} already includes slope angle. It should be noted that Equation (9.47) is not appropriate if the surface roughness is not reasonably smooth (i.e., approximately that of RCC).

The wavefront velocities are quite high, particularly for cases having the largest incident H_{m0} and associated larger values of H_{rms} on the landward-side slope. The primary cause for the high velocity is the wave forward momentum that is carried across

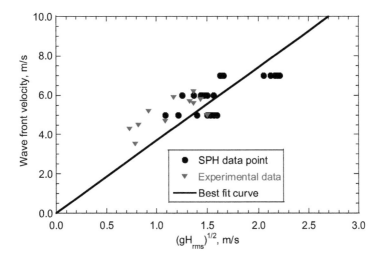

Figure 9.26 Estimated wavefront velocity determined from S2 to S6. (Adapted from Rao et al. (2012). Reproduced with permission from Elsevier.)

the crest of the levee by the combined wave and surge overtopping. Perhaps the wetted landward-side slope acts similar to a lubrication layer causing reduced frictional resistance to balance the wave momentum.

Chapter 10

Numerical study of turbulence overtopping and erosion

The high turbulence of the overtopping flow is an important factor for soil erosion and it can be responsible for the destruction of levees during combined wave over-topping and storm surge overflow. The goal of this chapter is to investigate the turbulence and erosion characteristics of HPTRM under combined wave overtopping and storm surge overflow. A three-dimensional (3D) hydrodynamic and sediment transport model, Estuarine and Coastal Ocean Model–Sediment transport (ECOMSED), has been calibrated and verified with overflow discharge and full-scale overtopping experimental results. This chapter presents overtopping hydrodynamic flow, turbulent shear stress, turbulent kinetic energy, and erosion rate at the toe of landward-side slope under combined overtopping condition. New equations for estimating turbulent shear stress, turbulent kinetic energy, and erosion rate at the toe of the landward-side slope were developed. The equivalency of erosion rate under storm surge overflow, combined wave, and surge overtopping were provided.

Numerical models are often employed to fill the gaps that physical models have due to the limitations of instruments. In recent years, the increase in computational power has enabled researchers to develop complicated numerical models to simulate the hydrodynamics of wave overtopping. Many models have been developed to simulate wave run-up and wave overtopping based on the nonlinear shallow water equations (Titov and Synolakis 1995; Lin and Liu 1999; Hubbard and Dodd 2002; Reeve et al. 2008; Yuan et al. 2014, 2015b). Lin and Liu (1999) employed a two-dimensional (2D) numerical model, which was a combination of a modified version of RIPPLE and a k-ε model to study wave overtopping above a seawall protected by a porous armor layer. Reeve et al. (2008) studied the effects of combined wave overtopping and storm surge overflow discharge on an earthen levee in a numerical flume by applying a modified model of Lin and Liu (1999). It expanded this model to three dimensions to solve complex free surface (Lin and Xu 2006). Reeve et al. (2008) used experimental data results from Soliman and Reeve (2004) to verify the numerical model, and developed empirical overtopping discharge formulas for combined wave overtopping and storm surge overflow. While Reeve et al. (2008) studied the combined waver overtopping and storm surge overflow, there is very limited information about the effect of combined wave overtopping and storm surge overflow on the HPTRM-strengthened levee on the toe of landward-side slope in particular.

Princeton Ocean Model (POM) is a widely used coastal ocean model (Blumberg and Mellor 1987; Ezer et al. 2003, 2008). A wetting and drying (WAD) scheme has been implemented into the POM (named POM-WAD) to simulate flow in near-coast regions where WAD processes prevail (Oey 2005). ECOMSED was developed from POM to simulate cohesive sediment resuspension, settling, and consolidation in shallow waters (e.g., Shrestha et al. 2000). The ECOMSED model solves equations for the conservation of mass, momentum, temperature, salinity, turbulent energy, macroscale turbulence, and sediment concentration. It incorporates the Mellor–Yamada Level 2.5 turbulence closure scheme, and an orthogonal curvilinear horizontal and σ-level vertical computational grid (e.g., Harris et al. 2005).

10.1 Numerical methodology

10.1.1 Governing equations

The Navier–Stokes equations under the assumption of hydrostatic pressure and Boussinesq approximation after Reynolds averaging in the Cartesian coordinate are used in ECOMSED for continuity equation and momentum equations. These equations are listed below.

$$\frac{\partial u}{\partial x} + \frac{\partial v}{\partial y} + \frac{\partial w}{\partial z} = 0 \tag{10.1}$$

$$\frac{\partial u}{\partial t} + u\frac{\partial u}{\partial x} + v\frac{\partial u}{\partial y} + w\frac{\partial u}{\partial z} = -g\frac{\partial \zeta}{\partial x} + \frac{\partial}{\partial x}\left(\varepsilon_h \frac{\partial u}{\partial x}\right) + \frac{\partial}{\partial y}\left(\varepsilon_h \frac{\partial u}{\partial y}\right) + \frac{\partial}{\partial z}\left(\varepsilon_z \frac{\partial u}{\partial z}\right) \tag{10.2}$$

$$\frac{\partial v}{\partial t} + u\frac{\partial v}{\partial x} + v\frac{\partial v}{\partial y} + w\frac{\partial v}{\partial z} = -g\frac{\partial \zeta}{\partial y} + \frac{\partial}{\partial x}\left(\varepsilon_h \frac{\partial v}{\partial x}\right) + \frac{\partial}{\partial y}\left(\varepsilon_h \frac{\partial v}{\partial y}\right) + \frac{\partial}{\partial z}\left(\varepsilon_z \frac{\partial v}{\partial z}\right) \tag{10.3}$$

$$\frac{\partial P}{\partial z} = -\rho g \tag{10.4}$$

where t is time; u, v, and w are the flow velocity components in the x, y, and z directions, respectively; ζ is the sea level; z is the vertical coordinate increasing upward with $z=0$ located at the undisturbed water surface and positive upward; P is the water pressure; ρ is the water density; g is the gravitational acceleration; ε_h and ε_z are the eddy viscosity of turbulent flow in the horizontal and vertical directions, respectively (Yuan et al. 2014).

The Mellor–Yamada Level 2.5 turbulence closure model (Mellor and Yamada 1982) and a prognostic equation (Mellor et al. 1998) for the turbulent turbulence macroscale are used to calculate the vertical eddy viscosity and diffusivity. The Mellor–Yamada Level 2.5 turbulence closure includes two partial differential equations to compute the turbulent kinetic energy (q^2) and a turbulence macroscale (l). The equation for the turbulent kinetic energy (without considering the variation of flow density) is:

$$\frac{\partial q^2}{\partial t} + u\frac{\partial q^2}{\partial x} + v\frac{\partial q^2}{\partial y} + w\frac{\partial q^2}{\partial z} = \frac{\partial}{\partial x}\left(\varepsilon_q \frac{\partial q^2}{\partial x}\right) + \frac{\partial}{\partial y}\left(\varepsilon_q \frac{\partial q^2}{\partial y}\right) + \frac{\partial}{\partial z}\left(\varepsilon_q \frac{\partial q^2}{\partial z}\right) + 2\left(P_s - \frac{q^3}{B_1 l}\right) \tag{10.5}$$

and the equation for the turbulence macroscale is:

$$\frac{\partial q^2 l}{\partial t}+u\frac{\partial q^2 l}{\partial x}+v\frac{\partial q^2 l}{\partial y}+w\frac{\partial q^2 l}{\partial z}=\frac{\partial}{\partial x}\left(\varepsilon_q\frac{\partial q^2 l}{\partial x}\right)+\frac{\partial}{\partial y}\left(\varepsilon_q\frac{\partial q^2 l}{\partial y}\right)+\frac{\partial}{\partial z}\left(\varepsilon_q\frac{\partial q^2 l}{\partial z}\right)$$
$$+lE_1 P_s-\frac{q^3}{B_1}\left[1+E_2\left(\frac{l}{\kappa L}\right)^2\right] \tag{10.6}$$

where P_s is the shear production, defined as $P_s=\varepsilon_z\left(\frac{\partial u}{\partial z}\right)^2+\varepsilon_z\left(\frac{\partial v}{\partial z}\right)^2$, $q^3/B_1 l$ is the turbulent dissipation; L is defined as $(\zeta-z)^{-1}+(H'+z)^{-1}$; H' is the average water depth at mean water level; ε_q $(=q l \chi_q)$ is the eddy diffusion coefficient for turbulence energy; ε_z is the vertical eddy viscosity; constant E_1 is 1.8; constant E_2 is 1.33; constant χ_q is 0.2; and κ is the von Karman constant. The last term in Equation (10.6) accounts for the effects of solid walls and the free surfaces on the length scale (Mellor and Yamada 1982). The vertical eddy viscosity (ε_z) is defined as $\varepsilon_z=q l \chi_z$. The coefficients χ_z are the stability functions related to the Richardson number and given by the following equation:

$$\chi_z=\frac{A_2\left(1-6 A_1/B_1\right)}{1-3 A_2 G_H\left(6 A_1+B_2\right)} \tag{10.7}$$

where $G_H=0$ without considering the variation of flow density and the constants used in Equation (10.7) are $A_1=0.92$, $A_2=0.74$, $B_1=16.6$, and $B_2=10.1$ (Mellor and Yamada 1982).

The 3D advection-dispersion equation for the transport of sediment is:

$$\frac{\partial C}{\partial t}+\frac{\partial u C}{\partial x}+\frac{\partial v C}{\partial y}+\frac{\partial (w-w_s) C}{\partial z}=\frac{\partial}{\partial x}\left(\varepsilon_h\frac{\partial C}{\partial x}\right)+\frac{\partial}{\partial y}\left(\varepsilon_h\frac{\partial C}{\partial y}\right)+\frac{\partial}{\partial z}\left(\varepsilon_z\frac{\partial C}{\partial z}\right) \tag{10.8}$$

where C is the suspended sediment concentration, and w_s is the settling velocity of the sediment flocs.

The finite differencing scheme was applied by Blumberg and Mellor (1987) to solve Equations (10.1–10.6), and Equation (10.8). The orthogonal grid was applied to the horizontal plane. In the vertical direction, a sigma coordinate was applied to transform both the sea surface and the bed bottom into coordinate surfaces. More details about this process can be found in Blumberg and Mellor (1987).

10.1.2 Boundary condition

The models neglect the wind stress and define the boundary conditions at the free water surface with $z=\zeta(x, y)$ are defined as below:

$$\rho\varepsilon_z\left(\frac{\partial u}{\partial z},\frac{\partial v}{\partial z}\right)=(0,0) \tag{10.9}$$

$$w=u\frac{\partial\zeta}{\partial x}+v\frac{\partial\zeta}{\partial y}+\frac{\partial\zeta}{\partial t} \tag{10.10}$$

$$\varepsilon_z\frac{\partial C}{\partial z}=0 \tag{10.11}$$

where $\zeta(x, y)$ is the free water surface level. At the near-bed bottom boundary,

$$\rho \varepsilon_z \left(\frac{\partial u}{\partial z}, \frac{\partial v}{\partial z} \right) = \left(\tau_{bx}, \tau_{by} \right) \tag{10.12}$$

$$w_b = -u_b \frac{\partial H}{\partial x} - v_b \frac{\partial H}{\partial y} \tag{10.13}$$

$$\varepsilon_z \frac{\partial C}{\partial z} = Q_e - Q_d \tag{10.14}$$

where $H(x, y)$ is the levee topography, (τ_{bx}, τ_{by}) is the bed frictional stress, and the Q_e and Q_d are the erosion flux and the deposition flux per unit area, respectively, with units of g/cm^2/h. The bottom frictional stresses can be determined by the following equation:

$$\tau_b = \rho C_D |U_b| U_b \tag{10.15}$$

where C_D is the drag coefficient and U_b is the velocity in the grid point that is closest to the solid boundary. Stephan and Gutknecht (2002) defined the drag coefficient as:

$$C_D = \left(\frac{1}{\kappa} \ln \frac{D - y''}{y''} + 8.5 \right)^{-2} \tag{10.16}$$

where D is the water depth, κ is the von Karman constant (0.4), and y'' is the mean deflected plant height. The mean deflected plant height of HPTRM-strengthened levee under combined storm surge and wave overtopping is 1.25 cm (Yuan et al. 2014).

The erosion flux (Q_e) can be evaluated when the bed properties are relatively uniform over the water depth and the bed is consolidated as follows (Mehta 1984):

$$Q_e = \begin{cases} Q_{e0} \left(\dfrac{\tau_b}{\tau_c} - 1 \right) & \text{if } \tau_b > \tau_c \\ 0 & \text{if } \tau_b \leq \tau_c \end{cases} \tag{10.17}$$

where τ_c is critical shear stress and Q_{e0} is an empirical coefficient with units of g/cm^2/h. Q_{e0} is also referred to as surface erosion rate and it ranges from 0.0018 to 0.036 g/cm^2/h (Ji 2007). Both the critical shear stress and the empirical coefficient can be determined by the experimental erosion data shown in Chapter 6.

It is assumed that most particles settled in flocs as long as near-bed shear stresses are small enough to prevent the flocs to be broken up (Einstein and Krone 1962). The deposition flux (Q_d) is related to settling rate and it occurs due to the formation and breakup of flocs, and it can be calculated as follows:

$$Q_d = \begin{cases} w_s C \left(1 - \dfrac{\tau_b}{\tau_d} \right) & \text{if } \tau_b < \tau_d \\ 0 & \text{if } \tau_b \geq \tau_d \end{cases} \tag{10.18}$$

where τ_d is critical shear stress above which flocs break up and no longer settle, and 0.1 Pa is used as suggested by Krone (1962). w_s is the settling velocity that depends on the size of sediment flocs. w_s can be calculated using the following equation:

$$w_s = \alpha_s \left(CG\right)^\beta \tag{10.19}$$

where G is the water column shear stress, and constants α_s and β are 2.42 and 0.22, respectively (Burban et al. 1990).

The erosion of vegetation and/or HPTRM is a complicated process. There are three usual erosion estimates including the erosion rate proportional to flow velocity, shear stress, and flow energy. The first two estimates are more commonly used. According to experimental measurements, the erosion rate of levee under combined overtopping is proportional to the bottom shear stress. The bottom shear stress (τ) was estimated by considering the influence of turbulence (Equations (10.5–10.6)). Meanwhile, Q_{e0} was determined by the experimental results, which enabled the numerical results of the erosion rate to match with the physical experimental better.

Sensitivity analyses were conducted to determine the effects of the critical shear stress (τ_d) for flocs deposition on the sediment transport and bed surface erosion. It was determined that the modeling results of the bed surface erosion on the levee crest and the landward-side levee slope were not sensitive to the deposition parameter, τ_d. Small deposition occurred when low-velocity overtopping flow passed the levee surface or when the levee surface became dry under combined overtopping with a small surge height. There was not sufficient time for deposited clay flocs for consolidation or compaction. These deposited flocs would be rapidly resuspended when the next wave peak passed. Meanwhile, very low suspended sediment concentration was observed both in physical and numerical modeling. This means that deposited flocs on the bed surface during the wave trough were limited and it could be neglected (Rao et al. 2012a, 2012b). This was the reason why the deposition process had less influence on sediment transport and final bed surface erosion (Yuan et al. 2014).

10.1.3 Model setup

The conceptual model of a levee embankment strengthened by HPTRM on the crest and along the landward-side slope has the same geometry as the levee embankment built in the full-scale overtopping tests, as described in Chapter 6. The dimensions of the levee embankment in the conceptual model were 26.12 m long × 3.25 m high × 3.66 m wide. The toe of the sea-side slope from the upstream boundary was 39.8 m. The width of the levee crest along the flow direction was 2.57 m. The sea-side had a slope of 1V:4.25H and the landward-side slope was 1V:3H. To reduce the reflection of wave impact at the downstream boundary, a sufficiently large reservoir was used in the downstream boundary (Figure 10.1).

Initially, the upstream water was at the same level as the levee crest, and there was no water in the landward-side slope and the downstream reservoir. The wave was generated at 39.8 m away from the sea-side slope toe. Eleven cases of storm surge overflow with various surge heights (0.2–1.2 m) were simulated. Thirty cases of combined wave overtopping and storm surge overflow with different surge heights, significant wave heights, and peak wave periods were also investigated with the ECOMSED model.

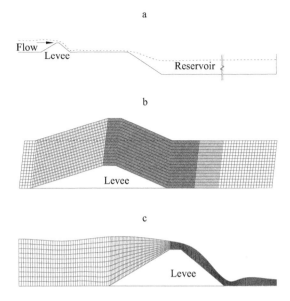

Figure 10.1 (a) Profile of levee embankment and downstream reservoir, (b) oblique view of the amplified horizontal orthogonal grid, and (c) amplified vertical sigma grid. (Adapted from Yuan et al. (2014). Reproduced with permission from the Coastal Education and Research Foundation.)

10.1.4 Numerical scheme

ECOMSED employs the finite differencing scheme to discretize the governing equations. An important advantage of the present model is the use of sigma coordination and a horizontal orthogonal curvilinear coordinate system. The σ-coordinate transformation, which was developed by Lu and Wai (1998), was applied to transform temporally and spatially. It has been confirmed as an efficient splitting method and a stable and effective approach by Chen et al. (1999). Grid sensitivity was analyzed by Li et al. (2012) and the same grid system was used for turbulence and erosion simulations. It had a grid size of 0.2 m in the lateral flow (y) direction (a total of 16 grids) and in the vertical flow (z) direction (a total of 10 grids). In the streamwise flow (x) direction, a grid size of 0.1 m with 25 grids on the crest and 96 grids on the landward-side slope was used (Figure 10.1).

10.1.5 Random wave generation

Random waves were applied as the upstream (sea-side) boundary condition for water level. The random waves were generated using the parameterized Joint North Sea Wave Project (JONSWAP) spectrum (Goda 1999) as:

$$S(f) = \frac{0.06238(1.094 - 0.01915\ln\gamma')}{0.230 + 0.0336\gamma' - 0.185(1.9 + \gamma')^{-1}} H_s^2 T_p^{-4} f^{-5} \exp\left(-1.25 T_p^{-4} f^{-4}\right) \gamma'^{\exp(-(T_p f - 1)^2 / 2\sigma^2)}$$

(10.20)

where $S(f)$ is the spectral density function, H_s is significant wave height ($= H_{1/3}$; defined as the average of highest 1/3 waves), T_p is the peak wave period, f is the wave frequency, γ' is the spectral enhancement factor, and the σ is 0.07 for $T_p f \le 1$ or 0.09 for $T_p f > 1$. The spectral enhancement parameter γ' ranges 1–6 and has a normal distribution with a mean of 3.3 and a standard deviation of 0.79 (Hasselmann 1973).

The wave linear superposition method was used to generate the random wave (Borgman 1969). The practical sea wave was assumed as the superposition of many random waves of different wave periods and different initial phases as:

$$\eta(t) = \sum_{i=1}^{M} \sqrt{2S(\hat{\omega}_i)\Delta\omega_i} \, \cos(\hat{\omega}_i t + \varepsilon_i) \tag{10.21}$$

where $\eta(t)$ is the water level, $\hat{\omega}_i$ is the representative frequency in the range of (ω_{i-1}, ω_i), and ε_i is the initial phase in the range of $(0, 2\pi)$. The range of spectrum circular frequency ω ($= 2\pi/f$) was decided by removing 2% of the total energy on both sides of the high and low frequency. The range of circular frequency was divided into M parts and $\Delta\omega_i = \omega_i - \omega_{i-1}$. Both $\hat{\omega}_i$ and ε_i were chosen randomly.

In wave overtopping studies, H_s is usually replaced with energy-based significant wave height (H_{m0}) (Hughes and Nadal 2009; Pan et al. 2013b). In this chapter, H_{m0} was also employed as the representative wave parameter and calculated by analyzing the generated random wave from Equation (10.16). The H_{m0} is defined as:

$$H_{m0} = 4.004\sqrt{m_0} \tag{10.22}$$

where m_0 is the 0-th spectral moment. The m_0 can be solved by:

$$m_0 = \int_0^{\infty} E(f) df \tag{10.23}$$

where $E(f)$ is the spectral energy density and f is the frequency. One example of the generated random wave with energy-based significant wave height (H_{m0}) of 0.778 m and peak wave period (T_p) of 7.0 s is shown in Figure 10.2.

10.2 Model calibration

ECOMSED hydraulic module, POM, was first calibrated with the experimental results for storm surge overflow and verified with empirical formulas. Then, the sediment module ECOMSED was calibrated with the full-scale overtopping experimental results. The experimental data from Chapter 6 was used to calibrate unknown parameters in the sediment transport model including the critical shear stress (τ_c) and the empirical coefficient (Q_{e0}). These sediment-related parameters can be obtained using Equation (10.17) with erosion flux given by the experiment data of Chapter 6. Bottom shear stress (τ_b) was predicted by Equations (10.15) and (10.16).

For storm surge overflow, the overflow discharge was estimated by broad-crested weir overflow equation (Henderson 1966) as:

$$q_s = 0.5443\sqrt{g}\left(-R_c\right)^{3/2} \tag{10.24}$$

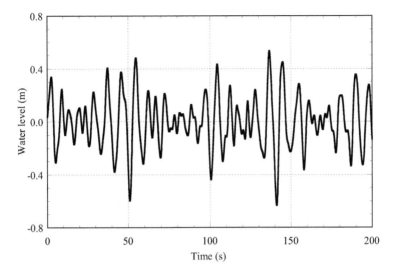

Figure 10.2 An example of generated random wave (JONSWAP spectrum, $H_{m0}=0.778$ m, $T_p=7.0$ s and mean sea level $=0$). (Adapted from Yuan et al. (2014). Reproduced with permission from the Coastal Education and Research Foundation.)

where q_s is the surge-only steady overflow discharge per unit length, g is the gravitational acceleration, and R_c is the freeboard, which is the difference between levee crest levee and still sea level. R_c can be negative for surge overflow.

Supercritical flow down the landward-side levee slope was estimated using the Chezy or Manning equations. The Manning equation for steady flow velocity (v_0) in the landward-side slope (in m/s) is given as follows:

$$v_0 = \left[\frac{\sqrt{\sin\theta}}{n} \right]^{3/5} q_s^{2/5} \tag{10.25}$$

where θ is the landward-side levee slope angle and n is the Manning coefficient. For grass-covered slopes exposed to steady supercritical overflow, Hewlett et al. (1987) recommended $n=0.02$ for slopes of 1V:3H. The bottom shear stress at the toe of the landward-side levee slope was estimated as:

$$\tau = \rho g \left(q_s / v_0 \right) \cdot i \tag{10.26}$$

where ρ is flow density, and i is the hydraulic gradient, which is the same as the levee slope when the uniform flow is formed near the toe of landward-side levee slope.

Eleven cases of storm surge overflow were simulated with the surge heights ranging from 0.2 to 1.2 m (Table 10.1). A comparison of modeling predictions from this study, experimental data of storm surge overflow, and an analytical solution for the storm surge overflow discharge is shown in Figure 10.3a. The round points are the experimental data measured in Chapter 6 and Chapter 7 in the full-scale storm surge overflow

Table 10.1 Surge height and erosion rate for storm surge overflow on the HPTRM-strengthened levee

Case number	Freeboard R_c (m)	Erosion rate E_s (mm/h)
1	−0.2	8.1040
2	−0.3	10.315
3	−0.4	12.651
4	−0.5	14.952
5	−0.6	16.883
6	−0.7	17.962
7	−0.8	19.124
8	−0.9	20.109
9	−1.0	21.613
10	−1.1	22.982
11	−1.2	24.121

Note: Negative R_c value means surge overflow.

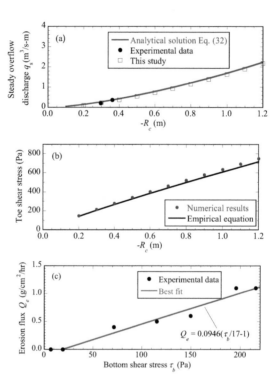

Figure 10.3 Model calibrated with empirical solutions and experimental data: (a) steady overflow discharge, (b) steady overflow toe shear stress, and (c) erosion flux Q_e at the surge height of 0.3 m. Experimental data were from full-scale overtopping tests in Chapters 6 and 7. Equation (10.24) was used to estimate steady overflow discharge and Equation (10.26) was used to estimated bottom shear stress at the toe of the landward-side slope. (Adapted from Yuan et al. (2014). Reproduced with permission from the Coastal Education and Research Foundation.)

experiments. The modeling predictions are in general agreement with experimental measurements. The numerical model results and laboratory data were slightly lower than the broad-weir equation. As shown in Figure 10.3b, the simulated shear stress at the toe of the landward-side slope was slightly higher than those predicted via the empirical Equation (10.26). The discrepancies were related to the increasing resistance caused by the grass and rough mat.

Figure 10.3c shows the measured erosion flux at the surge height of 0.3 m as a function of bottom shear stress. Four erosion data were measured at the same cross-section along the levee crest and the landward-side slope. Erosion data at the same cross-section were then averaged and erosion rate was measured in the experiments with units of cm/h. The density of the compacted clay was 2,000 kg/m^3. The erosion flux was calculated by multiplying the measured erosion rate by clay density with the unit of g/cm^2/h. The best-fit curve is given in the form of the Equation (10.27):

$$Q_e = 0.0946\left(\frac{\tau_b}{17} - 1\right) \tag{10.27}$$

where the critical shear stress (τ_c) and the empirical coefficient Q_{e0} for the HPTRM-strengthened levee should be 17 Pa and 0.0946 g/cm^2/h, respectively. The value of the best-fit Q_{e0} (0.0946 g/cm^2/h) was much higher than the typical value of 0.0018–0.036 g/cm^2/h (Ji 2007).

10.3 Storm surge overflow erosion

Eleven cases of storm surge overflow with surge heights ranging from 0.2 to 1.2 m were simulated. Simulations were performed to generate time sequences with a total duration of 200 s. Table 10.1 lists the input freeboard ($-R_c$) and the output erosion rate for each case.

Empirical correlations were sought to link the calculated erosion rate (E_s) at the toe of the landward-side slope to overflow parameters. The best correlation was observed between the maximum erosion rate and the negative freeboard (R_c) (Figure 10.4). The equation (Equation 10.28) illustrated by the solid curve in Figure 10.4 had a correlation coefficient of 0.904 and a root-mean-square (RMS) error of 0.408.

$$E_s = 13.81(-R_c)^{0.605} \tag{10.28}$$

where E_s is the erosion rate in mm/h at the toe of landward-side slope levee under storm surge. It should be noted that R_c must be entered as a negative number.

10.4 Combined wave/surge overtopping shear stress, turbulence, and erosion

The overtopping flow of combined wave and storm surge is much more complex than storm surge overflow. Flow down the landward-side slope caused by combined waves and surge overtopping is unsteady and more difficult to analyze. Thirty tests were run with freeboard (R_c) ranging from −0.3 to −0.9 m, and random waves with a JONSWAP spectrum and energy-based significant wave height (H_{m0}) ranging from 0.2 to 1.8 m.

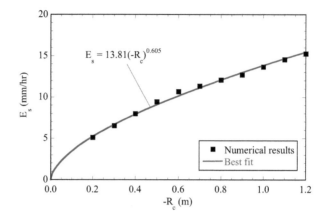

Figure 10.4 Erosion rate at the toe of the landward-side slope as a function of the negative freeboard in surge overflow. (Adapted from Yuan et al. (2014). Reproduced with permission from the Coastal Education and Research Foundation.)

Table 10.2 lists all the input wave parameters for all the 30 cases. Each simulation was performed to generate time sequences with a total duration in excess of 700s (corresponding to approximately 100 waves).

10.4.1 Turbulent shear stress

Bed shear stress is a fundamental variable in flow studies to calculate the sediment transport, deposition, and bed change (Wilcock 1996). In this study, Equations (10.15) and (10.16) were used to estimate the bottom shear stress because both time series of flow thickness and thickness were available. Figure 10.5 shows an example of time series of bottom shear stress at the toe of the landward-side slope under the combined wave overtopping and storm surge overflow with surge height of 0.3 m, energy-based significant wave height of 0.778 m, and peak wave period of 7 s.

The average bottom shear stress (τ) was determined by averaging the time series of bottom shear stress starting with data point 30 s and continuing to the end of the time series. Figure 10.6 shows the comparison of measured shear stress and calculates shear stress $\tau/g/(-R_c)$ at three monitoring locations (P1, P2, and P3) on the levee. P1 was located on the levee crest, P2 was located at the intersection of crest and landward-side slope, and P3 was located at the landward-side slope in the experimental levee section. Calculated shear stress $\tau/g/(-R_c)$ at P1 and P2 were in good agreement with the measured shear stress. At P3, the calculated shear stress matched well with the measured shear stress when the relative freeboard R_c/H_{m0} was less than -0.6. However, the calculated shear stress was lower than the measured shear stress when the relative freeboard was higher than -0.6. This difference could be related to the highly turbulent vertical velocity and the difficulty in the measurement of vertical velocity (w) due to the floating grass and shallow water thickness on the landward-side slope.

Table 10.2 Hydrodynamic parameters for combined wave overtopping and storm surge overflow

Case number	Freeboard R_c (m)	Energy-based significant wave height H_{m0} (m)	Peak wave period T_p (s)	Relative freeboard $-R_c/H_{m0}$ (-)
1	−0.3	0.381	7	0.787
2	−0.3	0.528	7	0.569
3	−0.3	0.566	7	0.530
4	−0.3	0.641	7	0.468
5	−0.3	0.778	7	0.386
6	−0.3	0.897	7	0.334
7	−0.3	1.014	7	0.296
8	−0.3	1.143	7	0.262
9	−0.3	1.277	7	0.235
10	−0.3	1.570	7	0.191
11	−0.3	2.312	7	0.130
12	−0.3	3.132	7	0.096
13	−0.3	3.840	7	0.078
14	−0.3	0.778	7	0.386
15	−0.3	1.570	10	0.191
16	−0.3	0.778	10	0.771
17	−0.6	1.014	7	0.592
18	−0.6	1.570	7	0.382
19	−0.6	2.030	7	0.296
20	−0.6	2.312	7	0.259
21	−0.6	3.132	7	0.192
22	−0.6	3.840	7	0.156
23	−0.6	4.660	7	0.129
24	−0.6	1.143	7	0.787
25	−0.9	1.570	7	0.573
26	−0.9	2.312	7	0.389
27	−0.9	3.132	7	0.287
28	−0.9	3.840	7	0.234
29	−0.9	4.660	7	0.193
30	−0.9	0.381	7	0.787

Source: Adapted from Yuan et al. (2014). Reproduced with permission from the Coastal Education and Research Foundation.

Note: freeboard ($-R_c$) is surge height above the crest.

The toe of the landward-side levee slope is the location of the maximum velocity, maximum shear stress, maximum turbulence, and maximum erosion, where the head-cut is typically formed first (e.g., Powledge 1989b). Because of the limitation of experimental equipment and the uncertainty of scale analysis, the overtopping quantities at the toe of the landward-side slope cannot be determined well in combined wave and surge overtopping conditions (Hughes 2008).

Based on the numerical results of the calibrated ECOMSED model, empirical correlations were sought to link the calculated mean shear stresses at the toe of landward-side slope to the combined overtopping parameters. The best correlation was observed between the mean shear stress $\tau/g/(-R_c)$ and the relative freeboard R_c/H_{m0} (Figure 10.6a). This equation is defined as follows:

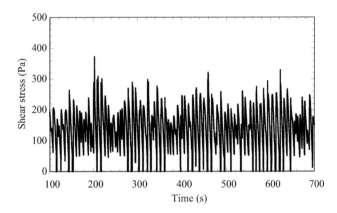

Figure 10.5 Time history of bottom shear stress at the toe of landward-side slope ($R_c = -0.3$ m, $H_{m0} = 0.778$ m, and $T_p = 7$ s). (Adapted from Yuan et al. (2014). Reproduced with permission from the Coastal Education and Research Foundation.)

$$\frac{\tau_m}{\rho g(-R_c)} = 0.1052\left(\frac{-R_c}{H_{m0}}\right)^{-0.282} \tag{10.29}$$

where τ_m is the time-averaged bottom shear stress at the landward-side slope during combined overtopping with the units of Pa, ρ is water density, and g is the gravity acceleration. The equation shown by the solid curve on Figure 10.7a had a correlation coefficient of 0.9991 and an RMS error of 5.744. In Equation (10.29), R_c must be entered as a negative number.

Figure 10.7b shows the comparison of numerical results of this study (Equation 10.29) with the empirical equations developed by Hughes and Nadal (2009). The numerical results were in good agreement with the estimates of Hughes and Nadal (2009).

10.4.2 Turbulent kinetic energy

The turbulent kinetic energy (k) (also represented by $q^2/2$ in the Equations (10.5) and (10.6)) was determined by averaging the time series of turbulent kinetic energy on the levee crest and on the landward-side slope starting with data point 30 s and continuing till the end of the time series. Figure 10.8 shows the comparison of measured turbulent kinetic energy ($k/g/(-R_c)$) and calculated turbulent kinetic energy at P2 and P3 as a function of relative freeboard. Calculated turbulent kinetic energy ($k/g/(-R_c)$) at P2 was in good agreement with the measured turbulent kinetic energy. At P3, the calculated turbulent kinetic energy matched well with the measured turbulent kinetic energy when the relative freeboard R_c/H_{m0} was less than -0.6. However, the calculated turbulent kinetic energy was lower than the measured turbulent when the relative freeboard was higher than -0.6. The difference was related to highly turbulent vertical velocity and the difficulty in the measurement of vertical velocity (w) because of the floating grass and shallow water depth on the landward-side slope.

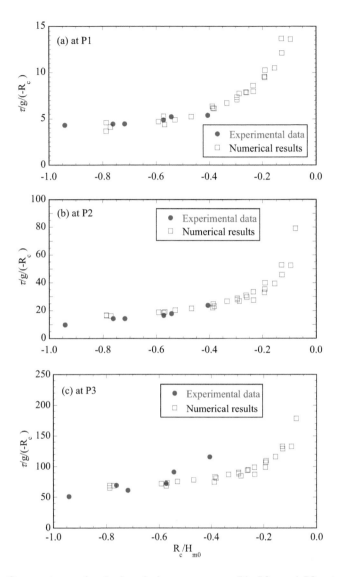

Figure 10.6 Comparison of calculated shear stress at P1, P2, and P3 with full-scale overtopping experimental results in Chapter 7. P1 was located at the levee crest, P2 was located at the intersection of the crest and landward-side slope, and P3 was located on the landward-side slope. (Adapted from Yuan et al. (2014). Reproduced with permission from the Coastal Education and Research Foundation.)

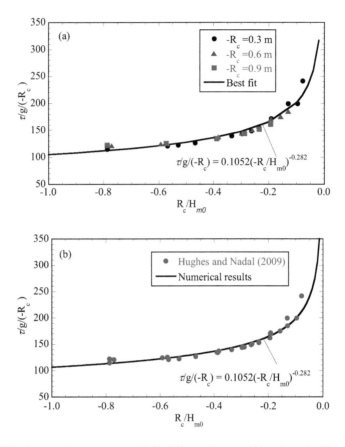

Figure 10.7 Turbulent shear stress ($\tau/g/(-R_c)$) at the toe of landward-side slope (a) as a function of relative freeboard (R_c/H_{m0}), and (b) compared to Hughes and Nadal (2009) empirical equations. (Adapted from Yuan et al. (2014). Reproduced with permission from the Coastal Education and Research Foundation.)

The best correlation was observed between the turbulent kinetic energy and the relative freeboard (Figure 10.9). This correlation equation is described as follows:

$$\frac{k}{g(-R_c)} = 0.83 \cdot \left(\frac{-R_c}{H_{m0}}\right)^{-0.285} \tag{10.30}$$

where k is the time-averaged turbulent kinetic energy at the toe of landward-side slope during combined overtopping with the units of m^2/s^2. The equation shown by the solid curve on Figure 10.9 had a correlation coefficient of 0.972 and an RMS error of 0.029.

10.4.3 Prediction of erosion rate at the toe of landward-side slope

Local clay erosion thickness varies periodically because local velocities vary periodically. Figure 10.10 shows an example of the time variation of the erosion thickness together

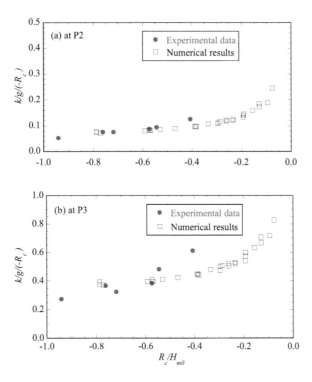

Figure 10.8 Comparison of calculated turbulent kinetic energy at P2 and P3 with the full-scale overtopping experimental data from Chapter 7. P2 was located at the intersection of the crest and the landward-side slope, and P3 was located on the landward-side slope. (Adapted from Yuan et al. (2014). Reproduced with permission from the Coastal Education and Research Foundation.)

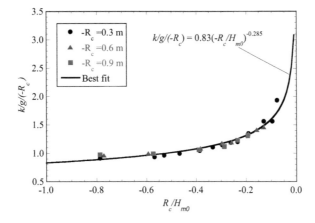

Figure 10.9 Turbulent kinetic energy ($k/g/(-R_c)$) at the toe of the landward-side slope as a function of relative freeboard (R_c/H_{m0}). (Adapted from Yuan et al. (2014). Reproduced with permission from the Coastal Education and Research Foundation.)

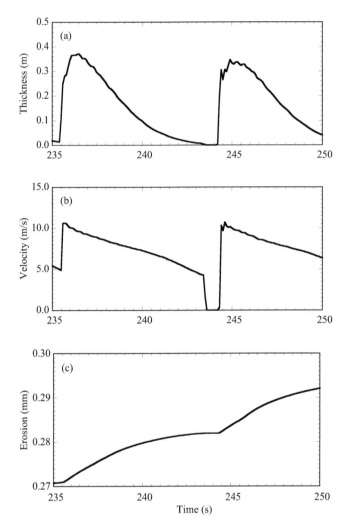

Figure 10.10 An example of time variation of (a) flow thickness, (b) flow velocity, and (c) erosion at the toe of the landward-side slope in two wave periods (surge height $R_c = -0.3$ m, energy-based significant wave height $H_{m0} = 1.277$ m, peak wave period $T_p = 7$ s). (Adapted from Yuan et al. (2014). Reproduced with permission from the Coastal Education and Research Foundation.)

with relative water thickness and flow velocity at the toe of the landward-side slope in two periods where freeboard (R_c) was −0.3 m and significant wave height (H_{m0}) was 1.277 m. As the wavefront flows over the toe, large-volume and fast-moving overtopping flow caused water thickness and flow velocity to increase rapidly. This resulted in the rapid erosion on the landward-side slope, whereas erosion rate slowed down as the wave trough reached to the toe and relative water thickness and flow velocity decreased to zero around the time point of 244 s.

Erosion rate was observed to be related to the relative freeboard with following equation (Figure 10.11a):

$$\frac{E}{-R_c} = 27.763 \left(\frac{-R_c}{H_{m0}} \right)^{-0.327} \tag{10.31}$$

where E is the erosion rate at the landward-side slope during combined overtopping with the units of mm/h. The equation shown by the solid curve on Figure 10.11a had a correlation coefficient of 0.9746 and an RMS error of 2.276.

Equation (10.31) can be written in terms of the steady overflow erosion rate (E_s) by substituting Equation (10.28) to yield:

$$E = E_s + 13.028 \left(-R_c \right)^{1.35} \cdot \left(\frac{-R_c}{H_{m0}} \right)^{-0.628} \tag{10.32}$$

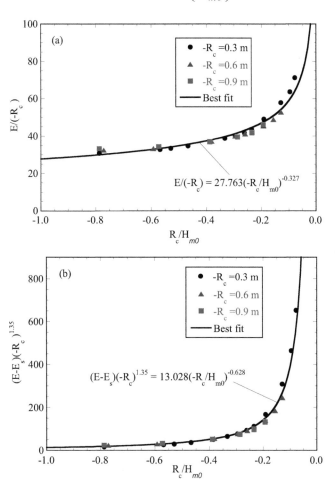

Figure 10.11 (a) Erosion rate and (b) equivalency of erosion rate at the toe of the landward-side slope as a function of relative freeboard (R_c/H_{m0}). (Adapted from Yuan et al. (2014). Reproduced with permission from the Coastal Education and Research Foundation.)

Figure 10.11b shows the equivalency of erosion rate at the toe of the landward-side slope as a function of relative freeboard. For higher values of significant wave height, the second term of Equation (10.32) dominates. Like any empirical equation, application of Equations (10.28–10.32) is limited to the range of tested parameters.

10.5 Erodibility and failure of HPTRM-strengthened levee under different overtopping conditions

10.5.1 Turf-element model and HPTRM-element model

Hoffmans et al. (2008) proposed a turf-element model to depict the protective mechanism of grass on levees against wave overtopping, in which the load was modeled by the uplift force caused by pressure fluctuations, whereas the strength of clay aggregate against the uplift force was characterized by the self-weight of the soil, cohesion, and grass strength. The maximum pressure peaks could be up to:

$$P_{max} = 18\tau_0 \qquad (10.33)$$

where P_{max} is the maximum pressure (N/m^2) and τ_0 is the bottom shear stress (N/m^2).

Incipient motion of a grass-clay aggregate occurs when the load is larger than the strength, thus the pressure should satisfy the following:

$$P \geq (\rho_s - \rho) g d_s + c_s + \sigma_g \qquad (10.34)$$

where P is the pressure fluctuation (N/m^2), ρ_s is the soil density (kg/m^3), ρ is the water density (kg/m^3), g is the gravity acceleration (m/s^2), d_s is the equivalent length of clay aggregate (m), $(\rho_s-\rho)gd_s$ is the self-weight, c_s is a characteristic cohesion (N/m^2), and σ_g is the grass strength (N/m^2).

The critical condition of lifting the grass-clay aggregate is reached if $P = P_{max}$ and if τ_0 is equal to the critical bed shear stress (τ_c), thus:

$$\tau_c = \alpha_\tau \left((\rho_s - \rho) g d_s + c_s + \sigma_g\right) \quad \text{with} \quad \alpha_\tau = 1/18 = 0.056 \qquad (10.35)$$

where α_τ is the constant coefficient with a value of 1/18.

In this chapter, the turf-element model was modified to account for the erodibility of the HPTRM-strengthened levee. The definitions of lift force, self-weight, and cohesion remained unchanged, but the grass strength (σ_g) was replaced with the grass-mat strength ($\sigma_{g, m}$):

$$\tau_c = \alpha_\tau \left((\rho_s - \rho) g d_s + c_s + \sigma_{g,m}\right) \qquad (10.36)$$

The direct replacement of the grass strength by the grass-mat strength is available. Mat applies force on the grass and not directly on the soil. Therefore, Hoffman et al. (2008) model is still available for the HPTRM-strengthened soil because only the value of grass force that acts on the soil changes. The d_s is the side length of the equivalent cube of a clay aggregate, which can be affected by a number of physical and chemical factors. Therefore, it is difficult to determine and can change depending on the clay characteristics.

The self-weight and cohesion are combined into one parameter defined as soil strength (σ_s) in this study. Thus, Equation (10.37) becomes the modified turf-element model for HPTRM-strengthened levee, which is called the HPTRM-element model in this study. Equation (10.37) is shown below:

$$\tau_c = \alpha_\tau \left(\sigma_s + \sigma_{g,m} \right) \tag{10.37}$$

10.5.2 Characterization of HPTRM-element model

Equation (10.37) depicts the relationship between the critical shear stress of HPTRM-strengthened clay and the properties of clay and HPTRM. The indicative values of the parameters in the HPTRM-element model could be calculated based on the EFA test results (Table 10.3). Table 10.3 shows that the critical shear stress (τ_c) and grass-mat strength $(\sigma_{g,m})$ of HPTRM-strengthened clay with good grass cover in EFA tests are consistent with the experimental results. Therefore, the erosion function obtained by the EFA tests can be used to predict the scour rate in the field, including scour on earthen levees against overtopping. The lower limit of HPTRM-good grass protected clay performance in the EFA tests was very close to its performance in the full-scale experiment. EFA tests can be a feasible method to predict the erosion rate – shear stress curves even when the physical shear stress is larger than the ones used in the EFA tests. However, the full-scale experiments are necessary to prove this.

The soil strength was small and negligible when compared to that of the grass-mat strength. The growth conditions of grass, roots in particular, can have a considerable effect on the performance of HPTRM. The strength of HPTRM with good grass cover was six times greater than that of with poor grass cover.

10.5.3 Failure process of HPTRM-strengthened levee against steady overtopping

The excess stress equation (a commonly used erosion model) was used to depict the erosion process and failure of clay with or without protection:

$$E = \frac{dh}{dt} = K_d \left(\tau_0 - \tau_c \right)^b \tag{10.38}$$

where E is the erosion rate (mm/h), h is the clay thickness (mm), t is the time (hour), K_d is the erodibility coefficient (mm/h/N/m^2), and b is an exponent.

Table 10.3 Indicative values for bare clay and clay with protections

Tests	Clay type	τ_c (N/m^2)	σ_s (N/m^2)	$\sigma_{g, m}$ (N/m^2)	K_d (mm/h/N/m^2)
EFA	Unprotected	0.059	1.05	0	5.263–19.231
	Poor grass	2.84	1.05	50	0.178–0.392
	Good grass	18.50	1.05	329	0.031–0.187
Full-scale	Good grass	17.00	1.05	303	0.028

Note: τ_c is the intercept on the x-axis in Figure 5.5b, d, and f. σ_s is assumed to be identical for all clays with protections. $\sigma_{g, m}$ is calculated by Equation (10.37). K_d represents the slopes of fitted lines in Figure 5.5b, d, and f.

The results of EFA tests were used as inputs in the Equation (10.38). Figure 5.5 shows that b is approximately equals to 1 for both the unprotected and HPTRM-strengthened clay and K_d is the slope of erosion function curve. The values of K_d in the EFA tests and the full-scale experiment are listed in Table 10.3 ($b = 1$) and shows that K_d in the experimental results is close to the lower limit of the EFA test results of HPTRM-strengthened clay. It should be noted that the erodibility coefficient (K_d) calculated under small shear stress can be used to predict large shear stress in levee overtopping conditions only when the linear relationship between erosion rate and shear stress exists. The results of EFA tests showed that with the increase of the shear stress, the erosion rate increased. It was also shown that this relationship was close to linear in the range of relative small shear stress, and it increased sharply after a certain threshold of shear stress was reached (Figure 5.3). However, HPTRM on the samples in EFA tests could not be anchored as in the field so they can easily be lifted up. Therefore, this chapter hypothesize that the linear relationship between erosion rate and shear stress may exist for much larger shear stress, if HPTRM is strictly fastened.

For the steady overtopping, the bottom shear stress on the levee slope was assumed to be constant over time. Then Equation (10.38) becomes:

$$\Delta h(t_0) = h(0) - h(t_0) = K_d(\tau_0 - \tau_c)t_0 \tag{10.39}$$

where $\Delta h(t_0)$ is the scour depth during t_0 (mm), $h(0)$ and $h(t_0)$ are the clay thickness at $t = 0$ and $t = t_0$, respectively.

In this chapter, the failure of HPTRM-strengthened clay was defined as the removal of an average of 100 mm thick soils from beneath the mats which was the average root depth of grass in the full-scale experiment and the EFA tests. Then, the relationships between the bottom shear stress and the duration for all samples calculated via Equation (10.39) were plotted in Figure 10.12. The lower limits of K_d shown in Table 10.3 were used during analyses.

It should be noted that the shear stress in overtopping for unprotected clay would cause unpredicted erosion when shear stress was much larger than the shear stress in EFA tests. It should also be pointed out that a linear relationship in case of small shear stress (<7 Pa) in EFA tests would no longer exist in overtopping condition. This is because clay can be eroded lump by lump, vertical headcut develops. Thus, the dotted curve was used in Figure 10.12 to emphasize that the erosion was unpredictable. It may be the same situation for the HPTRM-protected clay with inadequate grass. However, the HPTRM-protected clay with adequate grass can endure strong overtopping and high shear stress, and the linear relationship between shear stress and erosion rate still exists.

10.5.4 Failure process of HPTRM-strengthened levee against combined overtopping

It is known that on the landward-side slope, the sward contributes (albeit modestly) to the strength of the grass cover by covering the clay aggregates during overtopping, while the strength is dominated by the root and mat reinforcement and near the surface. The failure of HPTRM-strengthened levees during overtopping occurs in several stages. In the initial stage, elementary dispersed and loosened aggregates start being

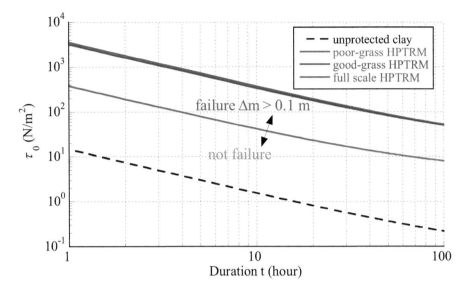

Figure 10.12 The relationships between bottom shear stress and duration. (Adapted from Yuan et al. (2015b). Reproduced with permission from the Coastal Education and Research Foundation.)

separated and torn out of the surface. Then, they are washed away by the flow. This process leads to many small shallow scours on the surface that can be gradually enlarged and deepened at an approximately uniform speed (Figure 10.13). However, no vertical headcut or pulling out of grass occurs because of HPTRM protection. When the scour depth is up to about the length of grass roots, the roots are completely exposed. Thus, they can no longer provide protection against flooding and the HPTRM fails. After that, the failure process may be the same as that for the earthen levees without protection and vertical headcuts lead to the final breach of the levees.

The bottom shear stress was assumed to be constant over time in Equation (10.39). Therefore, this equation was not applicable for the case of combined wave overtopping and storm surge overflow where the bottom shear stress on the landward-side slope changed over time due to the random waves. The erodibility coefficient (K_d) was assumed to be constant over time and b was assumed as one. Then, the Equation (10.38) becomes:

$$\Delta h(\Delta T) = h(0) - h(\Delta T) = K_d \sum_{i=1}^{n} \left[\int_{\Delta t_{1i}} \left(\tau_0(t) - \tau_c \right) dt \right] \qquad (10.40)$$

where ΔT is the total time, n is the total wave number, Δt_{1i} is the duration when the bottom shear stress is larger than the critical shear stress (τ_c) during ith wave passing, and $\tau_0(t)$ is the bottom shear stress on the landward-side slope which changes over time.

The time-averaged shear stresses are defined as:

$$\overline{\tau_0}_i = \int_{\Delta t_i} \tau_0(t) dt \bigg/ \Delta t_i \qquad (10.41)$$

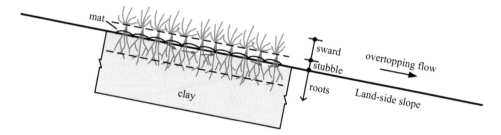

Figure 10.13 Shallow scours developing on the surface of the HPTRM-strengthened slope under overtopping flow. (Adapted from Yuan et al. (2015b). Reproduced with permission from the Coastal Education and Research Foundation.)

$$\overline{\tau_0}_{2i} = \int_{\Delta t_{2i}} \tau_0(t)\,dt \Big/ \Delta t_{2i} \tag{10.42}$$

$$\overline{\tau_0} = \int_{\Delta T} \tau_0(t)\,dt \Big/ \Delta T \tag{10.43}$$

$$\overline{\tau_0}_2 = \int_{\Delta T_2} \tau_0(t)\,dt \Big/ \Delta T_2 \tag{10.44}$$

where Δt_i is the duration during ith wave passing, Δt_{2i} is the duration when the bottom shear stress is smaller than the critical shear stress during ith wave passing, thus, $\Delta t_{2i} = \Delta t_i - \Delta t_{1i}$; ΔT_2 is the total duration when the bottom shear stress is smaller than the critical shear stress during ΔT, $\overline{\tau_0}_i$, $\overline{\tau_0}_{2i}$, $\overline{\tau_0}$, and $\overline{\tau_0}_2$ is the time-averaged shear stress during Δt_i, Δt_{2i}, ΔT, and ΔT_2, respectively.

Combining Equations (10.40) and (10.41) yields the scour depth in terms of time-averaged bottom shear stress:

$$\Delta h(\Delta T) = K_d \sum_{i=1}^{n} \left(\overline{\tau_0}_i - \tau_c\right)\Delta t_i - K_d \left[\sum_{i=1}^{n}\left(\overline{\tau_0}_i - \tau_c\right)\Delta t_i - \sum_{i=1}^{n}\left(\int_{\Delta t_{1i}} \left(\tau_0(t) - \tau_c\right)dt \right) \right] \tag{10.45}$$

which can be simplified as:

$$\Delta h(\Delta T) = K_d \sum_{i=1}^{n} \left[\left(\overline{\tau_0}_i - \tau_c\right)\Delta t_i + \left(\tau_c - \overline{\tau_0}_{2i}\right)\Delta t_{2i} \right] \tag{10.46}$$

It is known that $\overline{\tau_0}_i > \tau_c > \overline{\tau_0}_{2i}$ and $\Delta t_i > \Delta t_{2i}$. It was assumed that:

$$\overline{\tau_0}_i \gg \tau_c > \overline{\tau_0}_{2i} \tag{10.47}$$

Thus, Equation (10.41) can be approximated as:

$$\Delta h(\Delta T) \approx K_d \sum_{i=1}^{n} \left(\overline{\tau_0}_i - \tau_c\right)\Delta t_i = K_d \left(\overline{\tau_0} - \tau_c\right)\Delta T \tag{10.48}$$

When the Equation (10.47) is valid, Equation (10.48) can predict the scour depth of the landward-side slope of HPTRM-strengthened levees against the combined surge overflow and wave overtopping. It is difficult to measure $\tau_0(t)$ on the landward-side slope, especially at the toe where there is the maximum shear stress, turbulence, and erosion. More often, the time-averaged bottom shear stress $\overline{\tau_0}$ is available; therefore, the Equation (10.48) was used to replace the Equation (10.40).

This chapter used a 3D numerical model to simulate the HPTRM-strengthened levees against the combined overtopping with JONSWAP random spectrum. About 100 waves with different heights and freeboards were modeled in 700 seconds. An example of time series of the bottom shear stress at the toe of landward-side slope was plotted in Figure 10.14. Obviously, Equation (10.47) was satisfied in this case and $\Delta t_i \gg \Delta t_{2i}$.

An empirical equation for the time-averaged bottom shear stress at the toe of landward-side slope of a HPTRM-strengthened levee against the combined overtopping is presented as follows:

$$\frac{\overline{\tau_0}}{\rho g\left(-R_c\right)} = \alpha\left(\frac{-R_c}{H_{m0}}\right)^{-0.282} \quad \text{with} \quad \alpha = 0.1052 \qquad (10.49)$$

where ρ is the water density (kg/m^3), g is the gravity acceleration (m/s^2), R_c is the crest freeboard (m), and H_{m0} is the energy-based significant wave height (m). It should be noted that the application of Equation (10.49) is limited to the range of tested parameters used in Chapter 4, like the sea-side slope of 1V: 4.25H and the landward-side slope of 1V: 3H.

Combining Equations (10.48) and (10.49) yields the erosion process at the toe of the landward-side slope of HPTRM-strengthened levees before the failure of HPTRM:

$$\Delta h(\Delta T) = K_{d,H}\left[\alpha\rho g\left(-R_c\right)\left(\frac{-R_c}{H_{m0}}\right)^{-0.282} - \tau_{c,H}\right]\Delta T, \quad \Delta T \le \Delta T_0 \qquad (10.50)$$

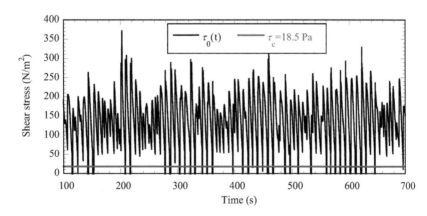

Figure 10.14 Time history of bottom shear stress at the toe of landward-side slope ($R_c = -0.3$ m, $H_{m0} = 0.778$ m, and $T_p = 7$ s). (Adapted from Yuan et al. (2015b). Reproduced with permission from the Coastal Education and Research Foundation.)

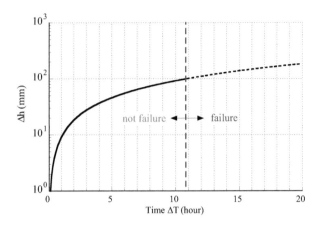

Figure 10.15 Time history of scour depth at the toe of the landward-side slope ($R_c = -0.3$ m, $H_{m0} = 0.3$ m, $\tau_{c,\,H} = 18.5$ Pa, and $K_{d,\,H} = 0.031$ mm/h/N/m^2). The dotted curve represents the unpredicted erosion process after HPTRM failure. (Adapted from Yuan et al. (2015b). Reproduced with permission from the Coastal Education and Research Foundation.)

where $K_{d,\,H}$ is the erodibility coefficients, $\tau_{c,\,H}$ is the critical shear stress of HPTRM-strengthened clay, and ΔT_0 is the duration before HPTRM failure, which can be calculated by the following equation:

$$\Delta T_0 = \frac{\Delta h\left(\Delta T_0\right)}{K_{d,H}\left[\alpha \rho g\left(-R_c\right)\left(\dfrac{-R_c}{H_{m0}}\right)^{-0.282} - \tau_{c,H}\right]} \qquad (10.51)$$

where $\Delta h(\Delta T_0)$ is the average length of grass roots, which is 100 mm in this study.

Figure 10.15 shows an example of the time history of scour depth at the toe of landward-side slope where $R_c = -0.3$ m, $H_{m0} = 0.3$ m, $K_{d,\,H} = 0.031$ mm/h/N/m^2, $\tau_{c,\,H} = 18.5$ N/m^2. It took approximately 10.8 hours for HPTRM to fail at the toe of landward-side slop with $\Delta h(\Delta T_0) = 100$ mm. Thereafter, much more severe scours occurred which may have resulted in vertical headcut. Figure 10.16 depicts the relationship between the overtopping conditions and the duration by using Equation (10.51) with $\Delta h(\Delta T_0) = 100$ mm. The duration could be determined by the freeboard and significant wave height specified by Equation (10.50) or Figure 10.16.

The time-averaged shear stress and velocity along the levee can be used to predict the erosion profile along the levee. Profiles of time-averaged shear stress and velocity distributions along the top and slope of levee against combined overtopping with different random waves are given in Figure 10.17a and b. Predicted erosion thickness along the levee after 6 hours overtopping was calculated by the Equation (10.48) (Figure 10.17c).

10.5.5 Discussion on validation of the proposed analytic equation

Validation of the proposed analytic update (Equation 10.48) was discussed in this chapter. Equation (10.48) exists only if $\overline{\tau_{0\,i}} \gg \tau_c$, or:

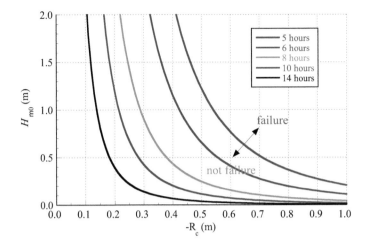

$$\sum_{i=1}^{n}\left(\overline{\tau_{0\,i}}-\tau_c\right)\Delta t_i \gg \sum_{i=1}^{n}\left(\tau_c-\overline{\tau_{0\,2i}}\right)\Delta t_{2i} \tag{10.52}$$

That is:

$$\left(\overline{\tau_0}-\tau_c\right)\Delta T \gg \left(\tau_c-\overline{\tau_{0\,2}}\right)\Delta T_2 \tag{10.53}$$

$\Delta T > \Delta T_2$. A sufficient but not necessary condition is:

$$\frac{\overline{\tau_0}-\tau_c}{\tau_c} \gg 1 \tag{10.54}$$

Figure 10.18 depicts the $\left(\overline{\tau_0}-\tau_c\right)/\tau_c$ at the landward-side slope of HPTRM-strengthened levees under different combined overtopping conditions and shows that $\left(\overline{\tau_0}-\tau_c\right)/\tau_c > 10$ and Equation (10.48) can exist under almost all conditions, except for weak overtopping (e.g., $-R_c < 0.1$ m). Fortunately, weak overtopping with small freeboard would not cause any significant damages to levees with strengthening systems.

This chapter established a 3D numerical model of full-scale levee against combined overtopping with freeboard (R_c) ranging from -0.3 to -0.9 m and random waves with a JONSWAP spectrum and energy-based significant wave height H_{m0} up to 4.66 m. The erosion rate was predicted by this model. The test results were used to compare with the estimates given by the Equation (10.48) and the shear stress ($\overline{\tau_0}$) estimated by the Equation (10.49) (Figure 10.19). A small overestimate (+20%) was observed which could be caused by the different conditions of mat and grass in the EFA tests and the full-scale tests. Thus, it was important to reduce the difference in samples to provide more precise predictions.

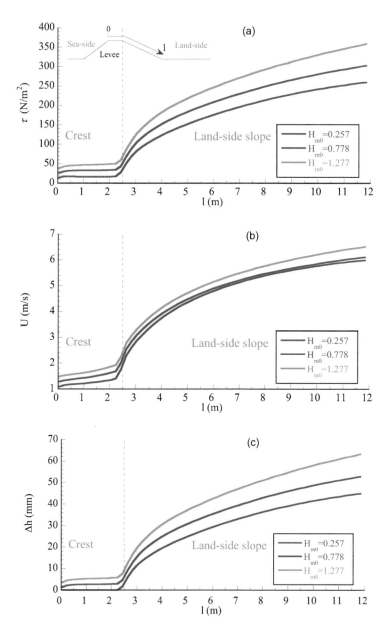

Figure 10.17 Profiles of time-average shear stress and velocity distributions along the top and slope of levee: (a) shear stress, (b) velocity, and (c) erosion thickness after 6 hours calculated by Equation (10.19) where $R_c = -0.3$m, $\tau_{c,\,H} = 18.5$ Pa, and $K_{d,\,H} = 0.031$ mm/h/N/m^2. (Adapted from Yuan et al. (2015b). Reproduced with permission from the Coastal Education and Research Foundation.)

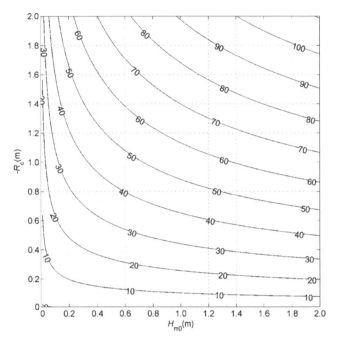

Figure 10.18 The values of $\left(\overline{\tau_0} - \tau_c\right)/\tau_c$ at the landward-side slope of HPTRM-strength-ened levees under different combined overtopping conditions ($\overline{\tau_0}$ is estimated by Equation (10.49), and $\tau_c = 18.5 \, \text{N/m}^2$). (Adapted from Yuan et al. (2015b). Reproduced with permission from the Coastal Education and Research Foundation.)

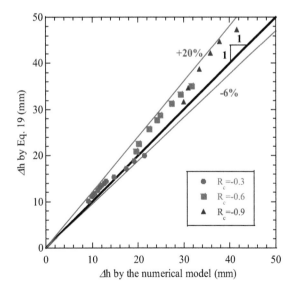

Figure 10.19 Validation of the proposed analytic equation ($\tau_{c,\,H} = 18.5$ Pa, and $K_{d,\,H} = 0.031$ mm/h/N/m^2). (Adapted from Yuan et al. (2015b). Reproduced with permission from the Coastal Education and Research Foundation.)

10.5.6 Discussion on linear relationship for HPTRM-strengthened clay

The hypothesis that allows to use EFA test results to predict the erosion in overtopping cases is that the linear relationship between erosion rate and shear stress exists even for much larger shear stress than those observed during EFA tests. There are two weak points of EFA tests. The first weak point is that EFA cannot create the bottom shear stress as large as the one on the landward-side slope with combined overtopping. There is very few equipment that can achieve this except the full-scale experiment. The second weak point of the EFA tests is that the HPTRM on the samples in EFA tests could not be anchored as in the field. Therefore, it is for this system to be lifted up. When HPTRM was lifted up, erosion rate would increase sharply like dotted curves shown in Figure 5.3. Thus, two series of tests would be used to prove this hypothesis, namely, EFA-like tests and full-scale tests.

Figure 10.20 shows the design of EFA-like tests. This equipment could create a big water head of about 2 m and offer the largest shear stress of 165 Pa at the test area. The grass and clay used in the tests were similar to those used in EFA tests. Unlike the EFA tests, the clay samples in this system were large enough (61 cm×20 cm×20 cm) for mat to be strictly fastened with a U pin as in the field. The edges of mat were fastened to the four sides of the cuboid sample. Six cases were conducted with three HPTRM-strengthened clay samples and two different shear stresses (132 and 165 Pa).

The experimental setup of full-scale tests is shown in Figure 4.3, with the wavemaker stopped. The steady storm surge overflow over the full-scale HPTRM-protected levee was run with the surge height $R_c = 30$ cm. After continuous overflow for 1 hour, erosion thicknesses of clay along the landward-side slope were measured. The shear stresses

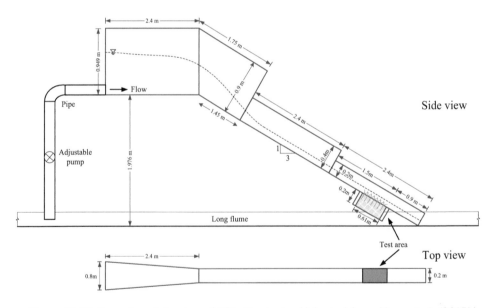

Figure 10.20 Experimental setup of EFA-like tests. (Adapted from Yuan et al. (2015b). Reproduced with permission from the Coastal Education and Research Foundation.)

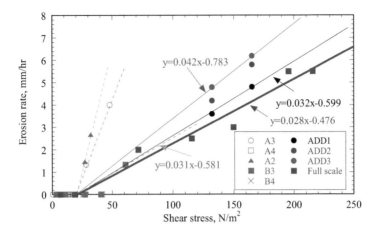

Figure 10.21 Comparisons of EFA test results for HPTRM-strengthened clays, EFA-like test results (named ADD1, ADD2, and ADD3), and full-scale test results (Adapted from Yuan et al. (2015b). Reproduced with permission from the Coastal Education and Research Foundation.)

along the slope were calculated by the numerical modeling. The largest shear stress was measured to be approximately 220 Pa.

Figure 10.21 shows the results of EFA-like tests and full-scale tests, together with EFA tests. No lift-up of mat was observed in both tests. Both showed a good linear relationship between erosion rate and shear stress even when the shear stress was large (165–220 Pa). The EFA-like tests predicted the erosion coefficient (K_d) at the ranges of 0.032–0.042 Pa. The low limit (0.032 Pa) was very close to the low limit erosion coefficient of EFA tests (0.031 Pa; Sample B3), and close to that of full-scale tests (0.028 Pa). Samples A2 and A3 showed a larger erosion coefficient possibly because of the relatively worse grass grow-conditions compared with Sample B3, EFA-like tests, and full-scale tests. Thus, choosing HPTRM-strengthened clay samples with better growth-condition grass and conducting more tests with mat that was strictly fastened in EFA tests should achieve better results. In addition, the low limit of the range of erosion coefficient should be used for erosion estimates of levee under combined overtopping.

References

Akkerman, G.J., Bernardini, P., van der Meer, J., Verheij, H., van Hoven, A. Field tests on sea defences subject to wave overtopping, *Proceedings of Costal Structures*, Venice, Italy, July 2–4, 2007.

American Society of Civil Engineers (ASCE). *Roller-Compacted Concrete*, Technical engineering and design guides no. 5, New York, 1994.

American Society of Civil Engineers/EWRI Task Committee on Dam/Levee Breaching (ASCE). Earthen embankment breaching. *Journal of Hydraulic Engineering*, 2011, 137(12): 1549–1564.

ASCE Hurricane Katrina External Review Panel. *The New Orleans Hurricane Protection System: What Went Wrong and Why?* Reston, VA: American Society of Civil Engineers, 2007.

Babaeyan-Koopaei, K., Ervine, D.A., Carling, P.A., Cao, Z. Velocity and turbulence measurements for two overbank flow events in River Severn. *Journal of Hydraulic Engineering*, 2002, 128(10): 891–900.

Baecher, G., Bensi, M., Reilly, A., Phillips, B., Link, L., Knight, S., Galloway, G. Resiliently Engineered flood and Hurricane infrastructure: Principles to guide the next generation of engineers. *The Bridge*, 2019, 49(2): 26–33.

Batchelor, G.K. *An Introduction to Fluid Mechanics.* Cambridge, UK: Cambridge University Press, 1974.

Besley, P. *Wave Overtopping of Seawalls. Design and Assessment Manual.* Wallingford, UK: HR-Wallingford, 1999.

Bingham, W.B., Schweiger, P.G., Holderbaum, R.E. Three innovative approaches to modify the spillways of existing embankment dams to accommodate larger floods using RCC, Modification of dams to accommodate major floods. *12th Annual USCOLD Lecture Series.* Fort Worth, TX, 1992.

Biron, P.M., Robson, C., Lapointe, M.F., Gaskin, S.J. Comparing different methods of bed shear stress estimates in simple and complex flow fields. *Earth Surface Processes and Landforms*, 2004, 29: 1403–1415.

Blanckaert, K., Lemmin, U. Means of noise reduction in acoustic turbulence measurements. *Journal of Hydraulic Research*, 2006, 44(1), 3–17.

Blumberg, A.F., Mellor, G.L. A description of a three-dimensional coastal ocean circulation model. *Coastal and Estuarine Sciences*, N. Heaps, ed., American Geophysical Union, Washington, DC, 1987, 4: 1–16.

Borgman, L.E. Ocean wave simulation for engineering design. *Journal of the Waterways and Harbors Division, ASCE*, 1969, 95(WW4): 557–583.

Bouws, E., Gunther, H., Rosenthal, W., Vincent, C.L. Similarity of the wind wave spectrum in finite depth water. Part I - Spectrum form. *Journal of Geophysical Research*, 1985, 90(C1): 975–986.

Briaud, J.L., Ting, F., Chen, H.C., Cao, Y., Han, S.W., Kwak, K. Erosion function apparatus for scour rate predictions. *Journal of Geotechnical and Geoenvironmental Engineering*, 2001, 127(2): 105–113.

Briaud, J.L., Chen, H.C. Levee erosion by overtopping during Hurricane Katrina. *Proceedings of the 3rd International Conference on Scour and Erosion*, London, UK, 2006.

Briaud, J.L., Chen, H.C., Govindasamy, A.V., Storesund, R. Levee erosion by overtopping in New Orleans during the Katrina Hurricane. *Journal of Geotechnical and Geoenvironmental Engineering*, 2008, 134(5): 618–632.

Briaud, J.L., Zenon, M.C., Stefan, H., Mark, E., Stacey. T. *Unknown Foundation Determination for Scour*, 2012, Texas Department of Transportation.

Burban, P.Y., Xu, Y., McNeil, Y., Lick, W. Settling speeds on flocs in fresh and sea waters. *Journal of Geophysics Research*, 1990, 95(C10): 18213–18220.

Carollo, F.G., Ferro, V., Termini, D. Flow resistance law in channels with flexible submerged vegetation. *Journal of Hydraulic Engineering*, 2005, 131(7), 554–564. DOI: 10.1061/(ASCE)0733-9429(2005)131:7(554).

Cea, L., Puertas, J., Pena, L. Velocity measurements on highly turbulent free surface flow using ADV. *Experiments in Fluids*, 2007, 42(3), 333–348.

Chen, X., Chiew, Y.M. Response of velocity and turbulence to sudden change of bed roughness in open-channel flow. *Journal of Hydraulic Engineering*, 2003, 129(1), 35–43.

Chen, Y.H., Cotton, G.K. *Design of Roadside Channels with Flexible Linings*. Washington, DC: United States Department of Transportation, 1988.

Chen, Y., Wai, W., Onyx, H., Li, Y., Lu, Q. Three-dimensional numerical modeling of cohesive sediment transport by tidal current in Pearl River Estuary. *International Journal of Sediment Research*, 1999, 14(2): 107–123.

Choi, Y., Hansen, K.D. RCC/Soil-cement: What's the difference? *Journal of Materials in Civil Engineering*, 2005, 17(4): 371–378.

Chow, V.T. *Open Channel Hydraulics*. New York City, NY: McGraw-Hill Book Company, , 1959.

Christensen, B.A. Open channel and sheet flow over flexible roughness. *Proceedings of the 21st International Association for Hydro-Environment Engineering and Research (IAHR) Congress*, Melbourne, Australia, 1985: 462–467.

Clopper, P.E. Protecting embankment dams with concrete block systems, *Hydro Review*, 1991, 10(2): 54–67.

Colagrossi, A., Landrini, M. Numerical simulation of interfacial flows by smoothed particle hydrodynamics. *Journal of Computational Physics*, 2003, 191: 448–475.

Dalrymple, R.A., Rogers, B.D. Numerical modeling of water waves with the SPH method. *Coastal Engineering*, 2006, 53: 141–147.

Dietrich, W.E., Whiting, P. Boundary shear stress and sediment transport in river meanders of sand and gravel. In *Water Resources Monograph*, R. Meandering, S. Ikeda, G. Parker, eds., 12. AGU, Washington, DC, 1989, 1–50.

Einstein, H.A., Krone, R.B. Experiments to determine modes of cohesive sediment transport in salt water. *Journal of Geophysics Research*, 1962, 67(4): 1451–1461.

EurOtop. *Manual on Wave Overtopping of Sea Defenses and Related Structures*. UK: Environmental Agency/Germany: German Coastal Engineering Research Council/the Netherlands: Rijkswaterstaat, Netherlands Expertise Network on Flood Protection, Netherlands, 2018.

Ezer, T., Oey, L.Y., Lee, H.C., Sturges, W. The variability of currents in the Yucatan Channel: Analysis of results from a numerical ocean model. *Journal of Geophysical Research*, 2003, 108(C1): 3012. DOI: 10.1029/2002JC001509.

Ezer, T., Hobbs, R., Oey, L.Y. On the movement of Beluga whales in Cook Inlet, Alaska: Simulations of tidal and environmental impacts using a hydrodynamic inundation model. *Oceanography*, 2008, 21(4): 14–23.

Finelli, C.M., Hart, D.D., Fonseca, D.M. Evaluating the spatial resolution of an acoustic Doppler velocimeter and the consequences for measuring near-bed flows. *Limnology Oceanography*, 1999, 44(7), 1793–1801.

Fuller, C. Concrete blocks gain acceptance as erosion-control systems, *Geotechnical Fabrics Report*, 1992, 10(3), Industrial Fabrics Association International, St. Paul, MN, 24–37.

Galperin, B., Kantha, L.H., Hassid, S., Rosati, A. A quasi-equilibrium turbulent energy model for geophysical flows. *Journal of Atmospheric Science*, 1988, 5(1): 55–62.

Garcia, C.M., Cantero, M.I., Nino, Y., Garcia, M.H. Turbulence measurements with acoustic Doppler velocimeters. *Journal of Hydraulic Engineering*, 2005, 131(12): 1062–1072.

Goda, Y. A comparative review on the functional forms of directional wave spectrum. *Coastal Engineering Journal*, 1999, 41(1): 1–20.

Goodrum, R. A Comparison of sustainability for three levee armoring alternatives. *Optimizing Sustainability Using Geosynthetics, the 24th Annual GRI conference Proceedings*, G.R. Koerner, R.M. Koerner, M.V. Ashley, G.Y. Hsuan, J.R. Koerner, eds., Industrial Fabrics Association International, Dallas, TX, March 16, 2011: 40–47.

Goring, D.G., Nikora, V.I. Despiking acoustic Doppler velocimeter. *Journal of Hydraulic Engineering*, 2002, 128(1): 117–126.

Grenzeback, L.R., Lukmann, A.T. *Case Study of the Transportation Sector's Response to and Recovery from Hurricanes Katrina and Rita*. Cambridge, MA: Cambridge Systematics, 2007.

Gudavalli, R., Ting, F.C.K., Briaud, J.L., Chen, H.C., Perugu, S., Wei, G. *Flume tests to study scour rate of cohesive soils*. Research Report Prepared for Texas Department of Transportation, Department of Civil Engineering, Texas A&M University, College Station, TX, 1997.

Gupta, R.S. *Hydrology and Hydraulic System*, 3rd Edition. Long Grove, IL: Waveland Press Inc., 2008.

Hahn, W., Hanson, G.J., Cook, K.R. Breach morphology observations of embankment overtopping tests. *Proceedings of the 2000 Joint Conference on Water Resources Engineering and Water Resources Planning and Management*, Minneapolis, MN, 2000: 1–10.

Hansen, K.D., Reinhardt, W.G. *Roller-Compacted Concrete Dams*. New York: McGraw-Hill, 1991.

Hansen, K.D. Erosion and abrasion resistance of soil-cement and roller-compacted concrete. *Research and Development Bulletin No. RD 126*, PCA, Skokie, IL, 2002.

Hanson, G.J., Cook, K.R., Britton, S.L. Observed erosion processes during embankment overtopping tests. *Proceedings of the 2003 ASAE International Annual Meeting*, Las Vegas, NV, 2003.

Hanson, G.J., Cook, K.R., Hunt, S.L. Physical modeling of overtopping erosion and breach formation of cohesive embankments. *Transactions of the ASAE*, 2005, 48(5): 1783–1794.

Harris, C.K., Traykovski, P.A., Geyer, W.R. Flood dispersal and deposition by near-bed gravitational sediment flows and oceanographic transport: A numerical modeling study of the Eel River shelf, northern California. *Journal of Geophysics Research*, 2005, 110(C9). DOI: 10.1029/2004JC002727.

Hasselmann, K. *Measurement of wind-waves and swell decay during the Joint North Sea Wave Project (JONSWAP)*. Hamburg, Germany: Deutsches Hydrographisches Institute, 1973.

Hedges, T.S., Reis, M.T., Owen, M.W. Random wave overtopping of simple sea walls: a new regression model. *Proceedings of the Institution of Civil Engineers – Water, Maritime and Energy*, 1998, 130(1): 1–10.

Henderson, F.M. *Open Channel Flow*. New York: MacMillian Publishing Co., Inc., 1966.

Hewlett, H.M., Boorman, L.A., Bramley, M.E. *Guide to the design of reinforced grass waterways*. CIRIA Report 116, London, UK, 1987.

Hoffmans, G., Akkerman, G.J., Verheij, H.J., Van Hoven, A., van der Meer, J.W. The erodibility of grassed inner dike slopes against wave overtopping. *Proceedings of the 31st International Conference of Coastal Engineering*, Hamburg, Germany, 2008: 2944–2956.

Hubbard, M.E., Dodd, N. A 2D numerical model of wave run-up and overtopping. *Coastal Engineering*, 2002, 47, 1–26.

Hughes, S.A. Levee overtopping design guidance: What we know and what we need. *Proceedings of the Solutions to Coastal Disasters Congress*, Turtle Bay, HI, 2008: 867–880.

Hughes, S.A., Nadal, N.C. Laboratory study of combined wave overtopping and storm surge overflow of a levee. *Coastal Engineering*, 2009, 56(3): 244–259.

Hughes, S.A., Scholl, B., Thornton, C. Wave overtopping hydraulic parameters on protected-side slopes. *Proceeding of 32nd USSD Annual Conference*, New Orleans, LA, April 23–25, 2012.

Huthnance, J.M., Humphery, J.D., Knight, P.J., Chatwin, P.G., Thomsen, L., White, M. Near-bed turbulence measurements, stress estimates and sediment mobility at the continental shelf edge. *Progress in Oceanography*, 2002, 52(2–4): 171–194.

IPCC. *Managing the Risks of Extreme Events and Disasters to Advance Climate Change Adaptation. A Special Report of Working Groups I and II of the Intergovernmental Panel on Climate Change.* Cambridge, UK: Cambridge University Press, 2012.

Ji, Z. *Hydrodynamics and Water Quality: Modeling Rivers, Lakes, and Estuaries.* Hoboken, NJ: John Wiley & Sons, Inc., 2007.

Johnson, F.A., Illes, P. A classification of dam failures. *International Water Power & Dam Construction*, 1976, 28(12): 43–45.

Kelley, D., Thompson, R. Comprehensive hurricane levee design: Development of the controlled overtopping levee design logic. *SAME Technology Transfer Conference and Lower Mississippi Regional Conference*, March 17–19, Vicksburg, MS, 2008.

Kim, S.C., Friedrichs, C.T., Maa, J.P., Wright, L.D. Estimating bottom stress in tidal boundary layer from acoustic Doppler velocimeter data. *Journal of Hydraulic Engineering*, 2002, 126(6): 399–406. DOI: 10.1061/(ASCE)0733-9429(2000)126:6(399).

Kindsvater, C.E. *Discharge Characteristics of Embankment-shaped Weirs.* Washington, DC: US Government Printing Office, 1964.

Koutsourals, M. A study of articulated concrete blocks designed to protect embankment dams, *Geotechnical Fabrics Report*, 1994, 12(7): 20–25, Industrial Fabrics Association International, St. Paul, MN.

Krone, R.B. *Flume studies of the transport of sediment in estuarial processes, final report.* Hydraulic Engineering Laboratory and Sanitary Engineering Research Laboratory, University of California, Berkeley, 1962, 110.

Lacey, R.W.J., Roy, A.G. Fine-scale characterization of the turbulent shear layer of an instream pebble cluster. *Journal of Hydraulic Engineering*, 2008, 134(7), 925–936.

Lancaster, T. A three phase turf reinforcement system a new method for developing geosynthetically reinforced vegetated linings for permanent channel protection. International Erosion Control Association, *Proceedings of Conference*, 1996, 27: 345–354.

Lane, S.N., Biron, P.M., Butler, J.B., Chandler, J.H., Crowell, M.D., McLelland, S.J., Richards, K.S., Roy, A.G. Three-dimensional measurement of river channel flow processes using acoustic Doppler velocimetry. *Earth Surface Processes and Landform*, 1998, 23: 1247–1267.

Li, L., Pan, Y., Amini, F., Kuang, C.P. Full scale laboratory study of combined wave and surge overtopping of a levee with RCC strengthening system. *Ocean Engineering*, 2012, 54(1): 70–86.

Li, L., Amini, F., Rao, X., Tang, H. SPH study of surge overflow and hydraulic erosion of earthen levee armored by articulated concrete blocks. *Current Development in Oceanography*, 2013, 6(2): 61–80.

Li, L., Amini, F., Pan, Y., Kuang, C.P., Briaud, J. Erosion resistance of HPTRM strengthened levee from combined wave and surge overtopping. *Journal of Geotechnical and Geological Engineering*, 2014, 32(4): 847–857.

Li, L., Yuan, S., Amini, F., Tang, H. Numerical study of combined wave overtopping and storm surge overflow of HPTRM strengthened levee. *Ocean Engineering*, 2015, 97: 1–11.

Lin, P., Liu, P.L.-F. Internal wave-maker for Navier–Stokes equations models. *Journal of Waterway, Port, Coastal, and Ocean Engineering*, 1999, 125(4): 207–215.

Lin, P., Xu, W. NEWFLUME: A numerical water flume for two-dimensional turbulent free surface flows. *Journal of Hydraulic Research*, 2006, 44(1): 79–93.

Lipscomb, C.M., Theisen, M., Thornton, C.I., Abt, S.R. Performance testing of vegetated systems and engineered vegetated systems. *Proceedings of the International Erosion Control Association Conference*, 2003, Number 34, January.

Liu, M.B., Liu, G.R., Lam, K.Y. Investigations into water mitigations using a meshless particle method. *Shock Waves*, 2002, 12: 181–195.

Lohrmann, A., Cabrera, R., Kraus, N.C. Acoustic-Doppler velocimeter (ADV) for laboratory use. *Proceedings of the Joint Conference on Fundamentals and Advancements in Hydraulic Measurements and Experiments*, 1994, Buffalo, New York, 351–365.

Lohrmann, A., Cabrera, R., Gelfenbaum, G., Haines, J. Direct measurements of Reynolds stress with an acoustic Doppler velocimeter. *Proceedings of the IEEE Fifth Working Conference on Current Measurement*, February 7-9, Institute of Electrical and Electronics Engineers, New York, 1995: 205–210.

Lu, Q.M., Wai, W.H. An efficient operator splitting scheme for three-dimensional hydrodynamic computations. *International Journal for Numerical Methods in Fluids*, 1998, 26: 771–789.

Lumley, J.L, Terray, E.A. Kinematics of turbulence convected by a random wave field. *Journal of Physical Oceanography*, 1983, 13(11), 2000–2007.

MacDonald, T.C., Langridge-Monopolis, J. Breaching characteristics of dam failures. *Journal of Hydraulic Engineering*, 1984, 110(5): 567–586.

Mansard, E.D., Funke, E.R. The measurement of incident and reflected spectra using a least square method. *International Conference on Coastal Engineering (ICCE)*, American Society of Civil Engineers, Washington, DC, 1980: 154–172.

McDonald, J.E., Curtis, N.F. *Applications of roller-compacted concrete in rehabilitation and replacement of hydraulic structures*. Technical Report REMR-CS-53, U.S. Army Engineer Waterways Experiment Station, Vicksburg, MS, 1997.

McLelland, S.J., Nicholas, A.P. A new method for evaluating errors in high-frequency ADV measurements. *Hydrological Processes*, 2000, 14, 351–366.

Mehta, A.J. Characterization of cohesive sediment properties and transport processes in estuaries, in Estuarine Cohesive Sediment Dynamics. *Lecture Notes on Coastal and Estuarine Studies*, Vol. 14, A.J. Mehta, ed., Springer, New York, 1984: 290–315.

Mellor, G.L., Yamada, T. Development of a turbulence closure model for geophysical fluid problems. *Reviews of Geophysics and Space Physics*, 1982, 20(4): 851–859.

Mellor, G.L., Oey, L.Y., Ezer, T. Sigma coordinate pressure gradient errors and the seamount problem. *Journal of Atmospheric and Oceanic Technology*, 1998, 15: 1122–1131.

Meselhe, E.A., Peeva, T., Muste, M. Large scale particle image velocimetry for low velocity and shallow water flows. *Journal of Hydraulic Engineering*, 2004, 130(9). DOI: 10.1061/(ASCE)0733–9429(2004)130:9(937).

Monaghan, J.J. Smoothed particle hydrodynamics. *Annual Review of Astronomy and Astrophysics*, 1992, 30: 543.

Monaghan, J.J., Kos, A. Solitary waves on a Cretan beach. *Journal of Waterway, Port, Coastal, and Ocean Engineering, ASCE*, 1999, 125: 145–154.

Monaghan, J.J. Smoothed particle hydrodynamics. *Reports on Progress in Physics*, 2005, 68: 1703–1759.

Monaghan, J.J. SPH simulations of shear flow. *Monthly Notices of the Royal Astronomical Society*, 2006, 365: 199–213.

Monin, A.S., Yaglom, A.M. *Statistical Fluid Mechanics: Mechanics of Turbulence*, Vol. 1. Cambridge, MA: MIT Press, 1971, 769.

Moody, L. F. Friction factors for pipe flow. *Transactions of the American Society of Mechanical Engineers,* 1944, 66.

Mori, N., Suzuki, T., Kakuno, S. Noise of acoustic Doppler velocimeter data in bubbly flow. *Journal of Engineering Mechanics*, 2007, 133(1): 122–125.

Moum, J.N., Gregg, M.C., Lien, R.C., Carr, M.E. Comparison of turbulence kinetic energy dissipation rate estimates from two ocean microstructure profilers. *Journal of Atmospheric and Oceanic Technology*, 1995, 12(2): 346–366.

Nadal, N.C., Hughes, S.A. Shear stress estimates for combined wave and surge overtopping at earthen levees. *Coastal and Hydraulics Engineering Technical Note ERDC/CHL CHETN-III-79*, U.S. Army Engineer Research and Development Center, Vicksburg, MS, 2009.

National Hurricane Center (NHC). *Costliest US Tropical Cyclones Tables Updated.* Miami, FL: NHC, 2018.

National Oceanic and Atmospheric Administrations/National Centers for Environmental Information (NOAA). *US Billion-Dollar Weather and Climate Disasters.* Asheville, NC: NOAA, 2019.

Nelson, R.J. Research quantifies performance of TRM reinforced vegetation. *Proceedings of the Sessions of the Geo-Frontiers 2005 Congress*, 2005, Austin, Texas, USA.

Nezu, I., Rodi, W. Open-channel flow measurements with a laser doppler anemometer. *Journal of Hydraulic Engineering*, 1986, 112(5). DOI: 10.1061/(ASCE)0733-9429(1986) 112:5(335).

Nikora, V.I., Goring, D.G. Flow turbulence over fixed and weakly mobile gravel beds. *Journal of Hydraulic Engineering*, 2000, 126(9), 679–690.

Nørgaard, J.Q., Andersen, T., Burcharth, H.F. Distribution of individual wave overtopping volumes in shallow water wave condition. *Coastal Engineering*, 2014, 83: 15–23.

Northcutt, P., McFalls, J. Performance testing of erosion control products – what have we learned after five complete evaluation cycles. International Erosion Control Association, *Proceedings of Conference*, 1998, 29: 199–218.

Nystrom, E.A., Oberg, K.A., Rehmann, C.R. Measurement of turbulence with acoustic Doppler current profilers – sources of error and laboratory results. In *Proceedings of the Hydraulic Measurements and Experimental Methods Conference*, Estes Park, CO, 2003, edited by T. Wahl, T. Vermeyen, K. Oberg, and C. Pugh. ASCE.

Oey, L.Y. A wetting and drying scheme for POM. *Ocean Modeling*, 2005, 9: 133–150.

Owen, M.W. *Design of Seawalls Allowing for Wave Overtopping.* Wallingford, UK: HR-Wallingford, 1980.

Pan, Y., Li, L., Amini, F., Kuang, C.P. Full scale HPTRM strengthened levee testing under combined wave and surge overtopping conditions: Overtopping hydraulics, shear stress and erosion analysis. *Journal of Coastal Research*, 2013, 29(1): 182–200.

Pan, Y., Kuang, C.P., Li, L., Amini, F. Full-scale laboratory study on distribution of individual wave overtopping volumes over a levee under negative freeboard. *Coastal Engineering*, 2015a, 97: 11–20.

Pan, Y., Li, L., Amini, F., Kuang, C.P. Overtopping erosion and failure mechanism of earthen levee strengthened by vegetated HPTRM system. *Ocean Engineering*, 2015b, 96: 139–148.

Pan, Y., Li, L., Amini, F., Kuang, C.P., Chen, Y.P. New understanding on the distribution of individual wave overtopping volumes over a levee under negative freeboard. *Journal of Coastal Research*, 2016, SI(75): 1207–1211.

Pan, Y., Chen, Y.P., Zhang, T.X., Hu, Y.Z., Yin, S., Yang, Y.B., Tan, H.M. Laboratory study on erosion of vegetated HPTRM system under high-speed open-channel flow. *Journal of Waterway, Port, Coastal, and Ocean Engineering*, 2018, 144(1): 04017038.

Perry, E.P. *Innovative Methods for Levee Rehabilitation*, Technical Report REMR-GT-26, U.S. Army Corps of Engineers, Waterways Experiment Station, 1998.

Pope, S. *Turbulent Flow.* Cambridge, UK: Cambridge University Press, 2000.

Powledge, G.R. Ralston, D.C. Miller, P. Chen, Y.H. Clopper, P.E. Temple, D.M. Mechanics of overflow erosion on embankments. I: Research activities. *Journal of Hydraulic Engineering*, 1989a, 115(8): 1040–1055.

Powledge, G.R. Ralston, D.C. Miller, P. Chen, Y.H. Clopper, P.E. Temple, D.M. Mechanics of overflow erosion on embankments. II: Hydraulic and design considerations. *Journal of Hydraulic Engineering*, 1989b, 115(8): 1056–1075.

Powledge, G.R., Pravdivets, Y.P. Overtopping of embankments to accommodate large flood events – An overview, Modification of dams to accommodate major floods. *12th Annual USCOLD Lecture Series*, Fortworth, TX, 1992.

Pullen, T., Allsop, N.W.H., Bruce, T., Kortenhaus, A., Schüttrumpf, H., van der Meer, J.W. *EurOtop: Wave Overtopping of Sea Defenses and Related Structures: Assessment Manual.* UK: Environment Agency/NL: Expertise Netwerk Waterkeren/DE: Kuratorium fur Forschung im Kusteningenieurwesen, Netherlands, 2007.

Ralston, D.C. Mechanics of embankment erosion during overflow. *Proceedings of the 1987 ASCE National Conference on Hydraulic Engineering*, Williamsburg, VA, 1987: 733–738.

Rao, X., Li, L, Amini, F., Tang, H. Numerical study of combined wave and surge overtopping over RCC strengthened levee system using smoothed particle hydrodynamics method. *Ocean Engineering*, 2012a, 54: 101–109.

Reeve, D.E., Soliman, A., Lin, P.Z. Numerical study of combined overflow and wave overtopping over a smooth impermeable seawall. *Coastal Engineering*, 2008, 55: 155–166.

Rune, S., Robert, G., Huang, Y. Simulated wave-induced erosion of the Mississippi River-Gulf Outlet levees during Hurricane Katrina. *Journal of Waterway, Port, Coastal, and Ocean Engineering, ASCE*, 2010, 136: 177–189.

Saucier, K.L. *No-slump Roller-compacted Concrete (RCC) for Use in Mass Concrete Construction*, Technical Report SL-84-17. Vicksburg, MS: U.S. Army Engineer Waterways Experiment Station, 1984.

Scawthorn, C., Porter, K. Enhancing resilience through risk-based design and benefit-cost analysis. *The Bridge*, 2019, 49(2): 16–25.

Schlicting, H. *Boundary Layer Theory*, 7th edition, McGraw-Hill: New York, 1987.

Schüttrumpf, H., Möller, J., Oumeraci, H., Grüne, J., Weissmann, R. Effects of natural sea states on wave overtopping of seadikes. *Proceedings of the 4th International Symposium on Ocean Wave Measurement and Analysis*, San Francisco, CA, 2001: 1565–1574.

Schüttrumpf, H., Oumeraci, H. Layer thicknesses and velocities of wave overtopping flow at seadikes. *Coastal Engineering*, 2005, 52: 473–495.

Shaw, W.J., Trowbridge, J.H. The direct estimation of near-bottom turbulent fluxes in the presence of energetic wave motions. *Journal of Atmospheric Ocean Technology*, 2001, 18(9): 1540–1557.

Shrestha, P.L., Blumberg, A., Di Toro, D., Hellweger, F. A three-dimensional model for cohesive sediment transport in shallow bays, Invited Paper, *ASCE Joint Conference on Water Resources Engineering and Water Resources Planning and Management*, July 30–August 2, Minneapolis, MN, 2000.

Sills, G.L., Vroman, N.D., Wahl, R.E., Schwanz, N.T. Overview of New Orleans levee failures: Lessons learned and their impact on national levee design and assessment. *Journal of Geotechnical and Geoenvironmental Engineering*, 2008, 134(5): 556–565.

Soliman, A., Reeve, D.E. Numerical study for small negative freeboard wave overtopping and overflow of sloping sea wall. *Proceedings of 3rd International Conference of Coastal Structures 2003, Portland, Oregon.* American Society of Civil Engineering, New York, 2004: 643–655.

Song, T., Chiew, Y.M. Turbulence measurement in nonuniform open-channel flow using acoustic Doppler velocimeter (ADV). *Journal of Engineering Mechanics*, 2001, 127(3), 219–232.

Soulsby, R.L., Dyer, K.R. The form of the near-bed velocity profile in a tidally accelerating flow. *Journal of Geophysical Research*, 1981, 86(C9): 8067–8074.

Soulsby, R.L. The bottom boundary-layer in shelf seas. *Physical Oceanography of Coastal and Shelf Areas*, B. Johns, ed., Elsevier, Amsterdam, 1983, 189–266.

Stapleton, K.R., Huntley, D.A. Seabed stress determination using the inertial dissipation method and the turbulent kinetic energy method. *Earth Surface Processes and Landforms*, 1995, 20(9): 807–815.

Stephan, U., Gutknecht, D. Hydraulic resistance of submerged flexible vegetation. *Journal of Hydrology*, 2002, 269(1–2): 27–43.

Sturm, T.W. *Open Channel Hydraulics*. New York, NY: McGraw-Hill Higher Education, 2001.

Svendsen, A., Putrevu, U. Nearshore mixing and dispersion. *Proceedings of the Royal Society of London. Series A: Mathematical and Physical Sciences*, 1994, 445(1925): 561–576.

Synder, W.H., Castro, I.P. Acoustic Doppler velocimeter evaluation in stratified towing tank. *Journal of Hydraulic Engineering*, 1999, 125(6), 595–603.

TAW. *Technical Report Wave Run-up and Wave Overtopping at Dikes*. Delft, The Netherlands: Technical Advisory Committee on Flood Defense, 2002.

Titov, V., Synolakis, C. Modeling of breaking and non-breaking long wave evolution and run-up using VTCS-2. *Journal of Waterway, Port, Coastal and Ocean Engineering*, 1995, 121: 308–316.

Tricito, H.M., Hotchkiss, R.H. Unobstructed and obstructed turbulent flow in gravel bed rivers. *Journal of Hydraulic Engineering*, 2005, 131(8), 635–645.

Trowbridge, J.H. On a technique for measurement of turbulent shear stress in the presence of surface waves. *Journal of Atmospheric and Ocean Technology*, 1998, 15: 290–298.

Trowbridge, J.H., Elgar, S. Turbulence measurements in the surf zone. *Journal of Physical Oceanography*, 2001, 31(8): 2403–2417.

Trowbridge, J.H., Elgar, S. Spatial scales of stress-carrying nearshore turbulence. *Journal of Physical Oceanography*, 2003, 33(5): 1122–1128.

U.S. Army Corps of Engineers (UACE), *Design and Construction of Levees*, Engineer Manual No. 1110-2-1913. Washington, DC: UACE, 2000.

U.S. Department of the Interior (USDOI). *Roller-Compacted Concrete*, 2nd Edition. Denver, CO: USDOI, 2017.

U.S. Department of the Interior. *Design and Construction Considerations for Hydraulic Structure: Roller-Compacted Concrete*, 2nd Ed., Bureau of Reclamation, Technical Service Center, Denver, CO, 2017.

van der Meer, J.W., Janssen, J. *Wave Run-up and Wave Overtopping at Dikes and Revetments*. Delft, The Netherlands: Delft hydraulics, 1994.

van der Meer, J.W. *Technical Report – Wave Run-up and Wave Overtopping at Dikes*. Delft, The Netherlands: Technical Advisory Committee for Flood De-fence in the Netherlands (TAW), 2002.

van der Meer, J.W., Janssen, J.P.F.M. Wave run-up and wave overtopping at dikes. *Wave Forces on Inclined and Vertical Structures*, N. Kobayashi, Z. Demirbilek, eds., ASCE, 1995, 1–27.

van der Meer, J.W., Schrijver, R., Hardeman, B., van Hoven, A., Verheij, H., Steendam, G.J. Guidance on erosion resistance of inner slopes of dikes from three years of testing with the Wave Overtopping Simulator. *Proceedings of ICE, Coasts, Marine Structures and Breakwaters*, Edinburgh, UK, 2009.

Victor, L., van der Meer, J.W., Troch, P. Probability distribution of individual wave overtopping volumes for smooth impermeable steep slopes with low crest freeboards. *Coastal Engineering*, 2012, 64: 87–101.

Villa, A. *Levee Armoring Workshop*, sponsored by US Army Corps of Engineers, New Orleans District, New Orleans, LA, 2007.

Voulgaris, B., Trowbridge, J.H. Evaluation of the acoustic Doppler velocimeter (ADV) for turbulence measurements. *Journal of Atmospheric and Oceanic Technologies*, 1998, 15: 272–289.

Wahl, T.L. Analyzing ADV data using WinADV. In Proceedings of the Joint Conference on Water Resources Engineering and Water Resources Planning & Management, Minneapolis, MN. July 30-August 2, 2000, 1–10.

Ward, D.L., Ahrens, J.P. *Overtopping Rates for Seawalls*. U.S. Army Engineer Waterways Experiment Station, US, 1992.

Wilcock, P.R. Estimating local bed shear stress from velocity observations. *Water Resource Research*, 1996, 32(11): 3361–3366.

Williams, J.J. Field observations and predictions of bed shear stresses and vertical suspended sediment concentration profiles in wave-current conditions. *Continental Shelf Research*, 1999, 19(4): 507–536.

Yuan, S., Li, L., Amini, F., Tang, H. Turbulence measurement of combined wave and surge overtopping over a full scale HPTRM strengthened levee. *Journal of Waterways, Coastal and Ocean Engineering*, ASCE, 2014, 140(4): 04014014.

Yuan, S., Tang, H., Li, L., Amini, F. Combined wave and surge overtopping erosion failure model of HPTRM levees: Accounting for grass-mat strength, *Ocean Engineering*, 2015, 109: 256–269.

Index

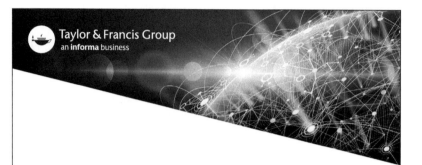

T - #0187 - 111024 - C234 - 246/174/11 - PB - 9780367535070 - Gloss Lamination